Dedication

To the memories of the brave Captain
Thomas F. Mantell USAF and Lt. Felix
Eugene Moncla Jr., USAF and the countless
other heroic military aviators who have
perished while intercepting UFOs.

Acknowledgements

The author would like to thank numerous people who have shared and discussed various reports of UFO phenomena with him. Most principally, Don and Vicki Ecker of the UFO Magazine, David Pfaff, Robert Morningstar, Linda Moulton Howe, Dr. Hal Puthoff, and the late Paul Murad, Leonard Stringfield, and Dr. Bruce Maccabee. Also, special thanks goes out to Ron Regehr, MUFON Chief of Documentation, for his critical reading of this manuscript.

Forward: "So Shall Your Seed Be"

And He brought Abraham forth outdoors and said, "Look now toward heaven and count the stars, if thou be able to number them." And He said unto him, "So shall thy seed be."

Genesis 15: 5

"Citizens of this Earth, we are not alone. God, in His infinite wisdom, has seen fit to populate His universe with other beings -- intelligent creatures such as ourselves."
John F. Kennedy from speech to be given in Dallas Texas
TRADE MART November 22, 1963 (UNDELIVERED)

"Until Change comes, something within us sleeps"

Dune, Frank Herbert

Abraham beneath the stars

President John F. Kennedy

This is not a book about "Rainbows and Unicorns". This is a book about the realities of the UFO phenomena and Coverup, and what it portends for human destiny in the Cosmos. Not all realities are pleasant, but they must be faced. The author loves tales of rainbows and unicorns, having raised three children, and I want those children to live in a Cosmos where we are part of a community of peoples, who like us, "have tears for things." To accomplish this goal certain basic truths must be accepted. We will journey through those truths in this book.

This book is an attempt to shape a future where humanity becomes part of a community of enlightened peoples in the stars. It is book about what will happen to our children, and their children after them. I want their future to be out among the stars. That means one must deal with hard questions now, so they will be spared those questions later. Other questions will concern them, of course, for every generation must face its difficulties. But this book is about addressing the hard questions of this generation.

In Military Intelligence, to understand a situation where much information is secret or fragmentary is described as "forming a situation estimate." Situation estimates are essential to military intelligence. When they are poorly done, things like Pearl Harbor and 9-11 happen. When they are done well, things like the victory at Midway or the end of the Cold War result. Based on the author's best analysis of scientific data, and most reliable reports, this book describes "reconstructions" of historical events and "Situation Estimates," of the realities humanity faces in the cosmos. Based on knowledges those realities, we can formulate, as a species, how to move forward towards our Cosmic destiny. Based on these knowledges, humanity will recognize it has been very fortunate in its evolving situation in the Cosmos, despite some problems that now must be overcome.

Based on the author's analyses, it now appears the Roswell incident, rather than being simply a curious news story of 1947, was instead a "hinge of history", a "watershed event", that has largely determined the realities, both open and secret, of the present day. In this book, will be presented an attempted reconstruction of what really happened in Roswell sometime around the early morning of the 4th of July in 1947, based on the best information and reports the author has been able to access. All around the Roswell incident exists a larger secret history, which this book will also explore.

Humanity is approaching a crucial crossroads in its history, it is a change of paradigm, a change of the fundamental set of premises that underly our definition of reality. We will understand soon, as a species, that we are not alone in the Universe. We will realize that we are part of the Cosmos and not separate from it; that the skies at night are a reflection of ourselves and we of them. With this change of consciousness everything will change, but at the same time nothing will change.

This book is an analysis of this transition in paradigms, that is occurring now. The analysis is based on two basic assumptions: A Supreme Being or Consciousness exists in the Universe, which links all living sentient beings; and two, a "Jungian Collective Subconscious", whose existence was first proposed by Carl Jung in 1912, which is a collective shared consciousness of all lesser sentient beings in the Cosmos. Drawing upon an excellent book of metaphysics by Cyd Robb, *A Simple Explanation of Absolutely Everything*, consciousness is fractal. With lesser consciousnesses being images of those more fundamental. Drawing on this model, the Jungian Collective Unconscious, is the result of the "quantum entanglement" of all lesser minds, born since the moment of creation. This Mind binds all humanity, and all sentient beings in the Cosmos, into something that is similar to the wave-particle duality of Quantum Mechanics—simultaneously, a relationship with a Supreme Being and a relationship linking all individual minds in in the Cosmos. Like electro-magnetism, it binds minds most closely those minds closest in space-time , and more subtly those minds far removed in space and time, so that the "here and now" affects it more deeply than things "in a galaxy far far away, and a long time ago".

By analogy, the Collective Unconscious, of say, a subset of the lesser sentient minds in the universe, all humanity on Earth, it can be thought of as the mind of a human in relationship to the individual nerve cells of the body. Or similarly, like the government of a nation, as opposed to a small village residing in that nation. Thus, it will be willing, in a situation of great danger, to sacrifice part of itself to save the whole. In this we confront the main truth of the existence of the Collective Unconscious, in that, in a violent and occasionally dangerous Universe, the Collective Unconscious, and its local genetic subsets, is concerned foremost with collective survival.

We will identify the Supreme Being or Consciousness with the God of the Bible in this tome. We will identify The God of Bible who exists and loves and helps humanity. His revealed word to us, particularly in the first Chapters of Genesis, is heavy with symbology and metaphor, and layers of meaning. What else could you expect from the Supreme Being of the Universe, Master of the Physics of the Big Bang and Quantum Entanglement, speaking to a bunch of shepherds? However, just as He expressed His love for us, He would, by extension, also love and help all sentient beings in the Cosmos. Also, by extension, the existence of a Jungian Collective Subconsciousness, not only links each human mind on Earth, but also other sentient minds not of this Earth. It can be conceived of as part of the Quantum Entanglement of all things, born as a singularity at the moment of the Big Bang. Based on these two premises, many mysteries can be understood, that we will encounter on the path of humanity to this crossroads. This Collective Unconscious is friendly to us over-all, so that in the end, we know we will join a community of enlightened species in the stars that surround us. That said, however, it must be realized that, as any astronomer will tell you, the Cosmos can be, in some space-time locales, a strange, violent and dangerous place. Thus, it is unrealistic to imagine all of this impending journey to a New Reality, this Cosmic Awakening for Humanity, will be entirely free from trauma and danger in some portions. However, the trend of the Cosmos can be seen through telescopes to be largely benign and favorable to life. So likewise, humanity must face this coming transition, with toughness, strength, and courage as well as hope and good will towards those fellow sentient beings living in the stars around us. But this Change must come, the "Sleeper must Awake," like the traumatic journey down the birth canal, it is the path of life. So, face the Change; it must come, and in the end it will be good.

A similar message is contained in the book *Allies of Humanity* by Marshall Vian Summers.

Humanity will be changed, because human beings are not just flesh and blood but a community of thoughts. The Heavens will change because we will become, consciously, part of them, and they with us. They themselves will be changed by our joining with them. Neither entity will be able to undo this event.

At the same instant, nothing will change, and God will remain the same, as will His love for us. His promises to us will remain unchanged. The Earth will remain the same, with its epic beauty and fullness. Humanity will be the same, with all its histories, strengths, and frailties intact. The Sun and stars will continue to burn brilliantly. The laws of physics that dictate this brilliance will be unchanged, as will be the principles that will allow humanity to sail the heavens, as others do.

This transition is like an awakening from a deep sleep, but it is also a journey.

The journey will be sometimes traumatic, as all great journeys are. It will be like child birth, an ordeal for both mother and child, but at the same time the path of life. This transition of consciousness is like awaking from a pleasant dream on a Monday morning, and then facing the day's realities. In your dream, time was an illusion, but upon awakening, you realize it is the first of the month, and rent is due. In your dream everyone you met was pleasant, but when you awaken you remember your beloved uncle liberated Death Camps at the end of the Second World War. In your dream, everyone was an invention of yourself, in your awakening everyone you deal with are others, with their own dreams.

The others we will learn to deal with, in the stars, will be very much like us in some ways, but dramatically different in others. We are like new graduates from the cosmic high school, which was Earth. We are thrust out of its sheltering halls, diploma in hand, to merge with a world of rugged adults. We must realize that high school graduation was only the 'end of the beginning' of our education, that we must now learn much more from the wider adult world, which is a Cosmos of unimaginable age in our reference. Humanity, used to being the rulers of all it surveyed, must now look up and realize they ruled only a speck of stardust, in a sea of stardust. Yet, we must also realize that those we encounter from the stars are, themselves, like us, merely specks of dust. They are lost in the sea of fate that is the Cosmos, just as we are.

As above, so below, as the saying goes. We will discover that we have been the beneficiaries of great kindness and protection by some the inhabitants of the stars around us, just as we will discover, also, that there are others who wish us great harm. We must show good will to all, returning kindness for kindness, and also to others who have not shown us kindness, an earnest desire for peace tempered by fortitude and courage. We are all equal in God's sight, and He watches over the stars as surely as He does the Earth. He rewards the just and kind, according to their works, and the brutal and cruel also, according to their works. It is the law of Cosmic Karma. This is a reminder to all who read this, from whatever star they call their Sun, as a worthy admonition, as one speck of star dust to another.

Therefore, let the journey begin. Let the Sleeper Awaken. Fear not, The God of the Whole Cosmos will be our traveling companion.

It will start by understanding how we arrived as a global civilization to this launchpad to the stars, how our heavenly vessel, both concrete and conceptual, came to be designed and constructed, and our mental journey will end in our destiny as an interstellar people. We will discover we are part of a Cosmic stream of life and sentience. We begin with our human self-concept from antiquity, arrive at the present view of mainstream science, and end with an extrapolation of all we know into the future. So let us begin our mental journey, our process of awakening. We will see it is but a small step for a human being, yet also a giant leap for humankind.

"Per Ardua, Ad Astra!" (by hard work, to the stars!)

There was a new Heavens and a new Earth, the old had passed away...

Rev 21: 1

Prologue: Rendezvous at Trinity

P-61s beginning a night patrol

As dusk gathered the P-80 Shooting Star jet fighters began to land and taxied away to their hangars. In the Western dark blue sky, the evening star, now visible, was shining brightly. So, it was reported. Taking their places on the long runways of Kirtland Army Air Field, the evening's flight of three jet-black P-61 Black Widow Night Fighters, with engines roaring, taxied into place. The P-61 was a radar equipped two engine propeller plane, a bloodied veteran of the just-ended Second World War. They were part of the 425th Night Fighter Squadron temporarily deployed to Kirtland from Hamilton Field in California, being recently recalled from Germany, still in ruins from the war. A P-61 , *Lady in the Dark*, was credited with the last air-to-air kill of that terrible war, now being forced into a fading memory by a world that wanted only peace. The Black Widow, so named because its glossy black paint made it resemble the venomous spider common in the Western US,   carried four 20 millimeter high velocity cannons on the bottom of its fuselage. Crowning the top of the Black Widow's fuselage, was a robotic tear drop shaped turret bearing four deadly 50 caliber machine guns. The crews wore red goggles to make their night vision as acute as possible after night fell and they would be patrolling the dark skies. Their night missions were called the Mars Patrols, after the Roman God of War.

It was the evening of the 3rd of July, the next day would see a riotous celebration across America of Independence Day and the great peace following the war that had ended just two years earlier. It was a war that had seen the deaths of 50 million human beings, many in Death Camps, and the devastation of large areas of Europe and East Asia. America had been spared the worst devastation and loss of life in that conflict, and this added to the sense of delirious celebration that 4th of July. Hot dogs, hamburgers, cold beer, and moments of reflection in shady back yards with family and friends, would be the order of the day. Though many would avoid thinking about the war on Independence Day, especially its last years, for some the day would be haunted by recurring nightmarish memories of past combat experiences. Added to this was the realization that the war had transformed the world into a new age, both wondrous and terrifying. It was a new age of missiles transiting space to strike targets hundreds of miles away, and nuclear weapons capable of leveling whole cities in an instant. So it was, that the Independence Day that would dawn the next morning, would be a day of mental escape, and a day of frightening specters to escape from.

But no such thoughts dominated the minds of the Black Widow pilots, as they the revved massive twin engines of their planes, and stood on their wheel brakes. The pilots and crew went over check lists carefully and exercised the control flaps and tail rudders of their planes, and finally the turret mounted 50 caliber guns, swinging them up and down as they rotated the turrets side to side. They made sure the ammunition feeds to them operated smoothly. The pilots looked to the South-East of Albuquerque, where a furious bank of thunderstorms raged in the desert distance, with the towering thunderheads pulsing and flickering with light as night deepened. But they were all World War Two combat veterans, and the storms that awaited them to the South East held no terror for them. The nose radars in their planes could see though the clouds and the Black Windows were so fast and rugged that neither storm lightnings nor winds could harm them. So they watched and soon the red flare from the tower was launched and the planes rocketed down the runaways to mount the darkened sky. Soon they were at altitude and racing on separate courses to a rendezvous at a point called Trinity to the South-East, to begin their all night patrol missions further to the South and East. They were going there to patrol the skies and protect the most vital military technology on Earth. Soon the lightning was flashing brilliantly off the glossy black wings, bursts of rain were covering the cockpit windows and the pilots and gunners took off their red goggles and put on eye patches so that bright lightning discharges would not temporarily disable their night vision in both eyes at once. They then armed their weapons. The storms held no terror for them, but something else did that night. A nameless terror, born a half century before in a fictional tale of heart rending destruction and horror, but whose fictional scenes had become the realities they had seen two years before. So, they flew on relentlessly, to the Trinity Rendezvous Point. They were jet-black engines of death, flying to defy that nameless terror and carry out their patrol mission that night. They were all veterans of a long war, now armed and airborne, life or death no longer mattered. To the crews of the Black Widows flying the Mars Patrol mission that night, the mission was the only thing that mattered.

Chapter 1 The Heavens and the Earth

"The Earth and its people are not an aberration on the Cosmos, they are a reflection of it."

David Pfaff

"In the beginning God Created the Heavens and the Earth..." Genesis 1: 1

Humanity, from its earliest recorded thoughts about itself, has never imagined itself either alone or monumentally important in the great scheme of things. We are said to have been made of the dust of the Earth. The word "human" comes from the same Indo-European root as the word "humus" or rich soil, dirt. Reality, was from antiquity, determined and explained to humanity by God or gods. In modern terms, God was the first ET (Extra-Terrestrial) entity humanity dealt with. Psychic contact with ETs, in some cases said to be non-corporeal, has been reported as some new and novel phenomena. However, it has always been present with humanity and when addressed to the Supreme Being, is called prayer. Despite being dirt, we were told we were "little lower than the angels" a nice thing to hear, considering angels are depicted as glorious beings and likened to the stars in the sky. These were the same skies humanity could clearly see utterly dwarfed them and all their works, even as an anthill is dwarfed by a person who steps over it. Yet, despite being a speck of dust, we were accorded importance in the holy writings of antiquity.

Modern biology still clings to the notion that life began on Earth with an ocean of "primordial soup" --water plus organic chemicals, that somehow organized itself into living cells.   This is a combination of water and organic chemicals is now known to exist everywhere in space. This avoids the direct question of whether life began on Earth, as it implies life being elsewhere in the stars, but this is only misdirection, for if life began here, then it surely began elsewhere by the same process. However, even a simple microbe is so complex that such life would take a long time to arise spontaneously. A competing hypothesis is that life was seeded here from space. This concept is called "Panspermia", and was originally proposed by the ancient Greek Philosopher Anaxagoras. According to him, life was everywhere in the universe. Panspermia is consistent with data that indicate life began very early on Earth, as soon as it had a liquid ocean, and on Mars as well. Panspermia is also consistent with images of microbe-like structures in carbon-rich meteorites such as Orguiel and Murchison.   This essentially says that life began in some "far away galaxy, a long time ago," and floated across space to Earth. This makes the question of life elsewhere in the stars, moot--it is there. However, either way it originated, life on Earth, and by extension, humanity, had humble origins. So both the Bible and biology say we were made of dirt, the common clay of existence, yet somehow, still a little lower than the angels.

This is vitally important concept, that the Creator and Ruler of the whole Universe has told you are a speck of dust, yet still important. As it is written in the Psalms 4: 3-5

> When I consider Your heavens, the work of Your fingers,
> The moon and the stars, which You have ordained,
> 4 What is man that You are mindful of him, And the son of man that You visit him?
> 5 For You have made him a little lower than the angels,
> And You have crowned him with glory and honor.

So, you are lower than the angels, but you are not nothing either, you are somewhere in between. This is mirrored in the saying of the Greek philosophers, that "man is the measure of all things." In the coming change of Cosmic consciousness and perception by humanity, this place in the Cosmos will not change, but instead will be understood more deeply. God fashioned us, either directly or indirectly. To the believer, we are here because God created us here, the details are not important, nor are stories told to us by strangers. The only thing that is important to believers in God, is that He is good and His will is fate, for us and everyone else. We are here in this place in the Cosmos because He has ordained it. We are here, on this planet, orbiting this star, on this galactic arm, it will be seen, right in the middle of things.

Our place in the Cosmos, it will be seen, is similar to the concept in physics of the "Mesoscale" of the atomic and nuclear physics scales that directly rule our existence. The Mesoscale (Middle scale) lies between the quantum Planck Scale of the very small and the Cosmic Scale of the very large. We, like the atoms, are dwarfed in size by the Cosmos we inhabit, and the mechanisms of our existence are dwarfed by the energies of the quantum Planck scale. "Why can't anything be simple," you ask, yet the essence of existence is complexity.

We learned how to think from the Greeks, it can be said, and learned how to pray from the Hebrews. The Greeks gave us philosophy and mathematics, beginning with Pythagoras, of the theorem, who studied in Egypt. Pythagoras invented the word "Cosmos" to describe the ordered universe, both Heavens and Earth. From the Hebrews, who also were students in Egypt, we learned that God, not your local despot, was the source of all law and that these laws were immutable. This concept was later adopted by the Greek philosopher Anaxagoras, who said that Divine Entity, whom he called "Nous", the Cosmic Mind, established all laws, physical and moral. Things became complicated in human affairs after this was all said.

Pythagoras, His Cosmos, and His Theorem

Adding to this vision of complexity, even the unseen portion of reality described, in the Western Tradition from the Bible, is itself complex. Like in other holy works from antiquity, the unseen heavens are full of life and sentience. Numerous types of angelic beings are described in the Bible: Cherubim, Seraphim, Archangels, and Angels (some fallen), are mentioned, but not explained in any detail. These beings are not flesh and blood, as ourselves, yet have their own dynamics and even wars. Dominions and organized powers, unseen but mighty, are discussed. So unseen reality is as complex as the seen, like the spectrum of radio and television broadcasts around us is complex, yet unseen unless one has the right receiver. To the mature physicist, humans possess only the senses that they needed for primitive survival. To sense radio or gravitational waves in antiquity would have been a harmful distraction. Yet modern humans remain limited in their perception of the Cosmos around us, despite our pretentions. To the serious physicist, much of what lies above and below our plane of reality is unknown, and that of which we do know is often baffling in its complexity. However, in the Western tradition, baffling complexity reaches its height in the dominant thought of Western religion--Judeo-Christianity.

In the land of Israel, in center of the ancient world, occupying the land bridge between the Africa of Egypt and the Eurasia of Babylon, stood the Jewish people, the Hebrews. They believed in One God, master of the Whole Cosmos, who had handed down laws for their behavior. Their central location and skills impressed the Greeks, the sometimes rulers and intellectuals of the ancient world, and affected their philosophy, as in the case of Anaxagoras. Likewise, the Jewish people learned both the Greek language, as a *lingua franca* of trade, and also the Greek mode of thought. So, a synthesis of Greek Philosophy and Hebrew Religion occurred, finally centered around one person, Jesus of Nazareth, a Jewish man, whom it was said was God, the Master of the Universe, incarnate. Nothing could confirm the basic message of the Bible that humanity is important in the great scheme of things than this concept of God choosing a human form to become incarnate in the physical universe. This was all written down in Greek, of course. With this complexity comes paradox.

Everything about Jesus and the belief structure that emerged surrounding him was paradoxical. He was not born in a palace but in a manger. He was not a great prince of a mighty people but instead raised in the household of a carpenter, in a nation subdued and down trodden by a mighty empire. He had no reputation as a scholar. He led no armies, fought no battles, and died an ignoble death. Yet is was said He rose from the grave and appeared to his now devastated disciples, who were then in hiding, and inspired them to fearlessly proclaim his teachings. Anyone who knows anything would say this story is most perplexing, and paradoxical. This gospel, or "Good News", was directed to be preached in all the Cosmos (so is the word in Greek) to every creature. This strange doctrine is that God Himself, knowing humanity was mired in misbehavior, and made of dust, took on this dusty form and reconciled, by his death, God with all flesh and blood. From the central location of Israel, and by the Hebrews' reputation as serious folk, this Gospel was spread throughout the Roman World of the time, where the concepts of the Greeks and Hebrews were well known. So, the history is deeply complex and puzzling, as is the belief structure that emerged from it. But everything pertaining to life is complex, and sometimes even paradoxical, and it is unrealistic to expect otherwise.

There is enough complexity in a single living cell as in all of human knowledge in all its libraries. So it is with the human place in the Cosmos. Nothing is simple nor to be understood in isolation: not a flower, not a star, not a human being. Nor is complexity limited to the physical. The universe runs on complex numbers, ask any engineer, and they, like ourselves, are partly real and partly imaginary. So it is with the human understanding of reality. To paraphrase Isaac Newton 'existence is made up of a small island of knowledge and vast shoreline of wonder'. Mathematical physics teaches us that in the midst of this complexity, however, there are constants that do not change even when a system undergoes dramatic changes. These invariants enable one to understand and predict the outcome of even dramatic processes on systems. Energy, momentum, symmetries, are termed "invariants" in physics, things that don't change. Sets of invariants are preserved in the collisions of high-energy particles as well as galaxies. So, it will be in the coming transition of human consciousness, from its present insular state in the Cosmos to final state of being joined to it. Two and two will still equal four after this happens, and humanities' position as 'little lower than the angels', somewhere between the Divine and microbes, will also be unchanged. But change must come, as it is also an invariant feature of reality.

Humanity in its history has undergone great changes of consciousness, some traumatic. The Copernican Revolution, a reemergence of the ancient understandings of the Greeks, removed Western Civilization and the Earth from being in the center of the Celestial map to just another point, just as Darwin removed humanity from being the pinnacle of nature to part of its fabric. Both ideas were embraced or rejected by various elements of the population. Yet, humanity preserved many invariants in these transitions, among them a reverence for laws and another, the habit of breaking of them, when it seemed convenient. This trait is reportedly not just the sole behavioral property of humans.

The universe and even the Earth considerably predates humanity, especially when one considers that the human species itself predates the invention of historical records, even those of dubious accuracy. Human beings, of the modern type, are estimated to have first appeared hundreds of thousands of years ago, yet presently accepted historical records, even of doubtful accuracy, begin only in the last six thousand years. Accounts from "prehistory", the hundred thousand years before civilized records – the writing down of old legends and 'oral histories,' describe contact not only with the Divine, and lesser non-corporeal beings, but also apparently flesh and blood beings from the heavens. These are the denizens of legend.

Early in the book of Genesis, what are described as "sons of God" found the daughters of Earth to be very attractive, and 'married' whomever they chose. The offspring of these unions are described as becoming 'great men' about whom 'many legends are written'. When this brief Biblical discussion is compared to the lively and detailed accounts from Greek Mythology of Zeus's and the lesser male gods sexual antics with human women, the parallels seem obvious. If we allow the certainty that we share the Cosmos with beings of like passions as ourselves, that is, flesh and blood ETs; ETs who predate us considerably in species origin and in the mastery of space travel, then these accounts could easily be mythologized versions of past contact. This is especially true, if we accept the accounts as basically possible, and therefore accept the possibility of some basically human ETs having visited here, and produced offspring with terrestrial women.

As unlikely as the possibility of the human species being genetically "trans-terrestrial" seems, according to modern biology, it must be remembered that modern biology also proposes that life began on Earth spontaneously from some primordial soup, a paradigm that appears increasingly dubious. Like all the sciences, biology is fundamentally based on phenomenology: what is observed. Accordingly, like all the sciences, biology must therefore evolve when confronted with new data. An avalanche of new data for human science is one of the many events that will accompany our change to a Cosmic- rather than Earth-centered consciousness. So despite protestations of biologists to these interpretations of ancient ET contact accounts, they must be entertained as basically factual. Once the concept of ET visitation by species far in advance of humanity in prehistory is considered, innumerable scenarios become plausible. Among them is the further possibility that some of these visitors may have been quite humanoid in behavior and genetics. The heroes of Greek Legend, Perseus, Thesus, ect..., men of remarkable abilities and intellect, all could be of mixed terrestrial and extraterrestrial heritage, and by extension all of humanity. Some Christian fundamentalists have told me that the 'sons of God' in Genesis were in fact, fallen angels, to which I replied: "their conduct seems hardly angelic."

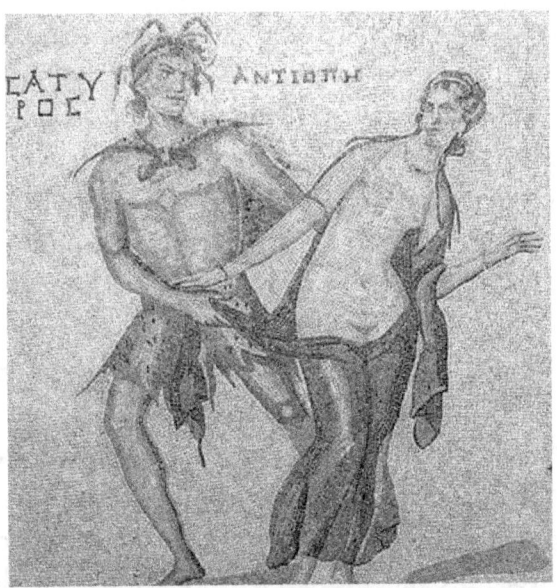

Zeus, as a satyr, seduces the Theban princess Antiope

Therefore, given the present advances of astronomy, with the discovery of the great age of the Cosmos, and hundreds of earthlike planets, and clouds of the basic chemicals of life floating in interstellar space, the reports from Genesis and elsewhere of the prehistorical ET visitation, must be considered basically plausible. So, also, is now plausible the concept that humanity, along with its misbehaviors, may be trans-terrestrial. This is the basic message to humanity from our own deep antiquity.

In the era of recorded history, which began with the invention of writing, both pictographic and phonetic, accounts of human encounters with non-terrestrial intelligences are frequent. The whole Bible is such an account. This statement speaks nothing against the validity of the Bible, but instead agrees with its depiction of a Cosmos that contains a complex hierarchy of sentient beings, with He whom we identify as the "Supreme Being" being the "Most High." To those who consider this vision of the Cosmos to be "too complex" and prefer the picture of universe inhabited by only "flesh and blood" intelligences like ourselves, I remind them of the statement of Richard Feynman, Nobel Prize Winner in Physics, who said, "Not everything that is true can be proven." In the author's experience as a physicist, much of importance cannot be measured directly, and the existence of some things can only be inferred from the measurable behavior of that which is known to exist. A working scientist actually resembles a person following a narrow trail through the deep woods most of the time, only daring to step off the beaten path on rare occasions. What lies in the deep woods far from the narrow trail is both unknown and barely comprehensible. Accordingly, profound statements concerning what is possible and impossible from scientists must be considered as more fanciful than factual. This is especially true now when we find ourselves on the shore of a vast mysterious ocean that we once thought was merely the remote and starry heavens, but will now become part of our world. Humanity has faced this situation before, however, and after some difficulties, has done well.

It is recorded the ancient Greek philosophers, to focus on the Western Tradition, were not content to merely philosophize, but also to carefully observe and calculate. Thus, by the 2nd century before Christ, the philosopher Aristarchus of Samos had deduced that the Earth was a sphere and moved about the Sun in a circular orbit like the other planets, and that the Moon was closer than the Sun and much smaller. He even believed, along with many other philosophers of his time, in "Many Worlds" beyond the Earth, each with their own inhabitants and also that the stars were other Suns with their own planetary systems. He finally journeyed to the Athens of Pericles and began to share these discoveries with those who would listen. However, the priesthood of the temples of Zeus and Mars, according to some accounts, had him promptly arrested and Pericles himself, had to come down to the jail and get him out and on the next ship out of the harbor. After this, the Spartans, utterly focused on war, conquered Athens and put an end to the discussions of the Cosmos by philosophers there.   Aristotle, court philosopher to Alexander the Great, a century later tried to speak about the Many Worlds philosophy of Democritus to his royal patron, Alexander, who, obviously not having a good day, reportedly broke down and wept, saying "I cannot even conquer this one world!"

Aristotle, noting this reaction of the "powers that be" and knowing where "his bread was buttered" in the court became much more conservative in his cosmology and more geocentric in his thinking after this. After all, the powerful on this planet desire their power to be great in their own eyes, and naturally view any talk of the universe where this world is not the most important one, to be annoying, if not subversive. This attitude was made more extreme by the later Romans, who said Rome was the immovable center of the universe and Rome was on Earth--end of discussion. Hence the popularity of the later philosopher Ptolemy with his geocentric cosmology with the Romans and the later Roman Church.

The Eastern limit of Christendom was Russia. In 1221, the Mongols, or Tartars, invaded and conquered Russia. In the 1230's they invaded Eastern Europe. They brought with them exploding iron bombs thrown by catapults. Europe had now been introduced to gunpowder.

The western limit of Christendom was Spain. In 711 Spain was invaded by the Muslim Moors, who conquered most of the region. However, aided by the rest of Christendom, the Spanish fought an 800-year war to drive out the Muslims, called "The Reconquista." This brutal conflict shaped Spanish society into a hardened Spartan-like organization, for whom war and religion were constantly intertwined. Conquest, followed by conversion of the peoples in captured lands, was a tactic formulated not only to add religious zeal to Spanish soldiers, but also to add legitimization to the aggressive waging of war. "We are furthering the kingdom of Christ" went the saying. Conversion by the sword to Catholic Christianity pacified the peoples conquered, who had previously been converted to the Muslim faith by the scimitar. In disputed areas where they could operate freely, Christian missionaries not only preached the Gospel but learned the local languages, and gathered intelligence for the soon-arriving Conquistadors. Later these new-found Catholics would be drafted into the armies of Catholic Spain, with abundant rewards, to further the advance of The Reconquista. This process became highly optimized. Finally, in 1492, the last Muslim stronghold in Spain fell to the Catholic Spanish. To the warrior aristocracy of Spain, peace led to the question of what to do with all these battle-hardened young men, who now had no more conquests to make, or captured lands to be distributed to them. Something would have be done, became the consensus among the rulers of Spain.

UFO in Renaissance Painting

In 1492, Columbus discovered the Americas for Spain. In 1519 Ferdinand Magellan, a Portuguese Captain, led and expedition for the Spanish around South America and across the Pacific to East Asia. He did not live to finish his expedition, but his crews did finally return to Spain in 1522, having circumnavigated the Earth for the first time. In 1521 Hernando Cortez and his conquistadors, aided by native American allies hoping for liberation, and the counsels of an Aztec princess who had previously been sold into slavery by her own people, conquered and destroyed the mighty Aztec empire. Despite the Spanish advantages of steel armor, gunpowder, and horses, all of which the Aztecs lacked, the Aztecs, called "Mexica" in their own language, put up savage, valiant, and well-organized resistance. It took a determined campaign of two years to conquer Mexico. Mexico held the premier indigenous civilization and military power of the "New World". Conquest of the rest of Latin America by the Spanish was a comparatively easy, yet brutal, task. The Spanish were then eager for more conquests. The procedures of conquest followed by conversion followed by recruitment of the converted into the Spanish armies, continued.

Conquistadors and allies versus Aztecs

In 1565, using expeditions from Mexico, the Spanish began conquering the Philippines, named after their king, Phillip. However, the Philippines held only primitive tribes scattered across many islands, and only meager amounts of gold. So, the Spanish were excited by tales of a rich kingdom to the North, with, it was said, much gold, and beautiful woman dressed in silks. It was also reported to have masses of heathen people, eager to have someone conquer them so they could become good Catholics. So, the Spanish conquistadors had dreams of sailing to North to claim that rich kingdom to the north, once the tiresome business of conquering the Philippines was concluded. This fabled kingdom to the north was known as Japan.

However, the Portuguese, then vigorous rivals to the Spanish, had arrived in Japan first. In 1543, the Portuguese began trade with the Japanese. The Japanese samurai warlords, Japan then being a disunited collection of warring fiefdoms, immediately learned how to make muskets and gunpowder from the Portuguese, who were eager to also sell these items. By 1556, the whole of Japan had samurai armies with battalions of skilled musketeers. Trouble soon followed these developments. The Portuguese fought the first naval battle with the Japanese in 1565 in Fukuda Bay.

The collision of these cultures, Portugal and Japan, could not have been more stark. The Portuguese, neighbors and students of the Spanish in all things, were a world ocean power. The Japanese did not know the world was round, had no compasses, almost no knowledge of regions outside of Korea and China, almost never sailed out of sight of land (except for samurai sea pirates) and had few boats bigger than a sampan. But the Japanese had a history of repelling brutal and powerful invasions from the sea--the Mongols. So, they came to look upon the Portuguese as of the same cloth. Thus, the Japanese samurai stood on the beach waiting to fight the Portuguese ocean-going galleons and their armored crews. The huge galleons, bristling with cannons, were full of steel-armored Portuguese Conquistadors full of religious zeal. To the samurai, the Portuguese galleons must have resembled alien starships. Yet the samurai faced them valiantly in any case, understanding that they were alien invaders. That was all the valiant samurai needed to know.

Mongols fighting Samurai, note exploding iron bomb

Meanwhile the Spanish finally conquered Manilla on Luzon in 1570 and made it the Capitol of the Philippines colony. A major clash between the Spanish and Japanese samurai occurred on Northen Luzon in 1582 as the Spanish consolidated their holdings in Luzon and encountered a Japanese base of samurai/pirates called Woukou. After a furious battle, the Spanish had victory, but in his report to Phillip, the King of Spain, the Governor General of the Philippines wrote in 1582

"The Japanese are the most warlike people in this region, and are well armed, having cannon, muskets and pikes, and wear body armor, which the wicked Portuguese have shown them how to make."

After this clash the Spanish concluded that Japan was too far away from Spain, too well armed, and too warlike to allow an armed conquest. Japan was not the Philippines, and it was not Mexico either.

The Spanish then tried to convert large numbers of Japanese peasants into good Catholics to achieve power by stealth. If Conquest and Conversion was not possible, perhaps Conversion followed by Conquest would be. This strategy had initial success, but the Tokugawa shogunate united Japan in 1603 and was annoyed that Christian areas in Southern Japan had resisted this unification. The Shogun finally outlawed Christianity as a foreign influence in 1639 and sealed off the country from all European contact. This period of Japanese isolation would last more than 200 years.

Poland, was the bastion of Western Civilization in Eastern Europe, facing the steppes of Russia, from whence came strange ideas and occasional raids from the Turks and Mongols from outside Christendom. Even Poland was electrified, however, first by reports and actual looted writings and artifacts from the Aztec civilization, paraded around Catholic Europe by the Spanish in the 1520's as spoils of war. In addition, Europe became awash with the lost learning of the Greeks, an occurrence that would help trigger the Renaissance in Italy. The Polish Catholic Church official, Copernicus, excited by these ideas, rediscovered the concepts of Aristarchus of Samos, and being wise as well as brilliant, arranged to have these ideas of a heliocentric Solar system, with a moving Earth, published after his death in 1543.

Nicholas Copernicus, Wary and Wise.

In 1561 war in the heavens was observed over Nurenburg Germany. This battle featured what appeared to be an hours-long clash in the sky, with ships exploding and crashing to Earth and burning. In 1604 a supernova occurred--the explosion of a known star. Europe was by that time well acquainted with gunpowder and things blowing up, sometimes by accident. So these events suggested to the learned that things were not only complex in the heavens, but that bad things happened there, just like on Earth.

These ideas were soon championed by Galileo and Kepler. Galileo even crafted a fine telescope, an instrument originally developed in the Netherlands for war, and observed the movement of the moons of Jupiter, like a miniature solar system of Copernicus. He published these observations in *Dialogue Concerning the Two Chief World Systems* in 1632. Galileo, however, was in the shadow of the Vatican, who had him arrested and threatened with torture if he did not recant. *"You only think you saw moons orbiting Jupiter, right?"* The Inquisition told him, he wisely agreed and they sent him home under house arrest for the rest of his life. The Pope considered himself, as "Chief Primate", to be master of the Earth, and saying publicly that the Earth was "just another planet" was naturally viewed as subversive. So, this view of the Cosmos and our place in it, as masters of the only place that matters, has a long tradition with the powerful here. So does the reaction of the powerful on Earth to those who suggest otherwise. Kepler was only slightly more fortunate.

Galileo Under House Arrest

Kepler was an astronomer making a living working in part of the national defense industry of those days. He was an astrologer tracking the movements of Mars, the planet of war. War, either waging it or preparing for it, was the preoccupation of governments in the late Middle Ages, just as now. Since the aristocracy believed in astrology, this meant Mars was the "money planet" for those who studied its movements for the courts of Europe. But Kepler knew, based on careful eye observation of its movements, that Mar's orbit was not the simple, circle-centered-on-the-Sun as defined by the Copernican Model. Mar's orbit was a strange oval around the Sun, not a circle.

Johannes Kepler

Using all the new tools of mathematical analysis of the late Middle ages, Arabic numerals, and the rediscovered mathematics of the ancient Greeks, Kepler, using Mar's orbit, deduced the laws of planetary motion. The planets did orbit the Sun, he confirmed, but in elliptical orbits, not strictly circular ones. The planets also sped up as they neared the Sun on these oval orbits, and slowed as they moved farther away. The planet's time to complete an orbit also was found to be proportional to the average radius of the orbit to the 3/2 power. For this reason, Earth would move in its orbit faster than Mars, and the two planets would approach each other every two years. At these close approaches called apparitions, Mars would dominate the night sky and its details would most visible through telescopes.

Kepler's work was an astonishing piece of mathematical and intellectual analysis. If Copernicus had fired the first shot of the Scientific Revolution, then Kepler began its first artillery barrage. This was published in his great work *Astronomy Nova*, in 1609. Still, many learned men clung to the reassuring model of Earth as the motionless center of the universe, around which everything else revolved. All this trouble stemmed from observations of Mars.

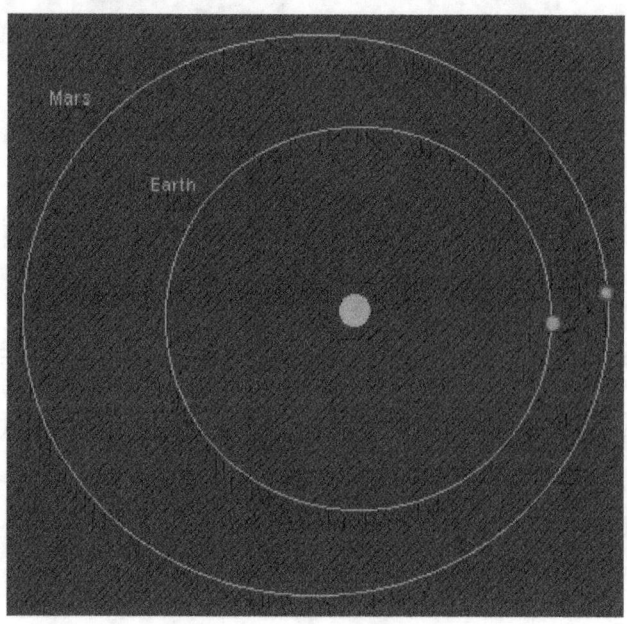

The Orbits of Earth and Mars

Chapter 2 The Age of Mars

The heavens declare the Glory of God, and the firmament showeth His handiwork, night after night showeth speech and day unto day showeth knowledge. Psalm 19: 1-2

Mars, the god of War

Vienna Austria was the gateway into Europe from the steppes of Russia. In the summer of 1683, in a third and most massive attempt, the Ottoman Turks attempted to conquer the city. It was the clash of civilizations on an epic scale. Vienna held out stubbornly until the Turkish armies were defeated before its gates by an allied army of Germans, Austrians, and Poles. Crucial to this victory was the charge of the Polish cavalry, experienced in fighting the Turks on the steppes. But what was also crucial was the Western possession of large numbers of good telescopes. This tool enabled long-distance sighting of artillery and monitoring of enemy's movements. The Turks could not make them, and so were deficient in both supplies of them and personnel skilled in their use. Use of the telescope, pioneered by Galileo for astronomy, was born because of its association with Mars, the god of war. However, in the aftermath of this victory, far from Vienna, another momentous event unfolded, in England.

The *Principia*, short for *The Mathematical Principles of Natural Philosophy, was* published by Isaac Newton on 5 July 1687, four years after the victory at Vienna. It laid out the laws of motion of matter, and the rules of calculus needed to analyze these motions, all this as preface in order to present the law and dynamics produced by Gravity. Newton's laws and gravitation verified and explained Kepler's laws of planetary motion. This was especially true of Mars, whose orbit was noticeably noncircular. This analysis also explained the very noncircular orbits of comets, formerly objects of terror in the sky.

Isacc Newton

The *Principia* has been described as the greatest work of science ever known, and triggered an explosion of Western science and technology. In one stroke it validated the Copernican Model of the Solar system. Its primary motivation had been to understand the laws that governed the heavens, but it did far more. With it, the intellectual fixation of the Western World became science, as opposed to religion. Therefore, the victory of the revolution of Copernicus in human thought, now gave rise to a revolution in science.

Before Newton, it can be said the humanity looked at the heavens with wonder and ascribed only to God any comprehension of its dynamics. After Newton, humanity looked on the heavens differently, as a puzzle that could be solved, understood, and exploited for good, using the concept of laws of physics that all matter, even in the heavens, obeyed. Humanity after Newton, it could be said, was still lower than the angels, *but we could now tug at their sleeves.*

This also raised greatly in the eyes of literate, the status of the new type of learned man, the man of science. It was just such a man, Benjamin Franklin, who discovered the link of electricity from a handheld piece of amber rubbed on wool, with the terror-inducing lightning of the clouds. Using his invention of the lightning rod, all the West no longer feared the thunderbolts of Zeus. He then lent his credibility to the Declaration of Independence signed July 4, 1776, just one day short of 199 years since the publication of Newton's *Principia*. Thus, a consequence of Newton's revolution in physics was a revolution in politics. If the planets were all equal in moving about the Sun, then so were people.

This was a difficult concept for the world to accept, being dominated by warring elites and religions, whose doctrines used the supposed inequalities of human beings to employ them for the maintenance and expansion of their despotisms. The New American Republic was free from rule by Britain, but also free from its protections. The Barbary Pirates, an annoyance to the major European powers, who found it easier to pay them "protection money" than try to eradicate them, began to prey on American shipping in the Mediterranean. Without a navy, we were an "easy mark." So, we paid them protection money like everyone else, knowing it was far cheaper than building a navy. Building a navy and fighting them was also discouraged by John Adams, the second US President, who noted the pirates were vassals of the Ottoman Turks and well placed in the Muslim world. Adams remarked, 'Once we start fighting them, we will have to fight them forever.' Thomas Jefferson, who followed him as President, felt differently. And so a navy was built, armed with the latest cannons and equipped with the latest telescopes, and set sail to war.

It was armed with these knowledges, and new technologies, that humanity once again began to study Mars, the planet of war. Mars was observed, especially on its close approaches to Earth, with telescopes of increasing power and precision. Unlike the Moon, which was apparently an airless grey desert, Mars, by the mid 1800's, was perceived to have polar caps and dust storms carried by winds across its surface. To the Western World, now the premier abode of science, it appeared Mars was Earthlike as opposed to Moonlike. Armed by the Many Worlds philosophy of the ancient Greeks and the confidence in the human intellect inspired by Newton, some boldly suggested that Mars was inhabited, like Earth.

This sentiment was compounded by the publication, by Charles Darwin, in November 24, 1859, of the book, *Origin of the Species*. If humanity has risen from the less intelligent animal kingdom by natural processes over time, became the question, could not a similar process have occurred on Mars? Equally important, was the publication, also in 1859, of the *Dynamical Theory of the Electromagnetic Field*, by James Clerk Maxwell, which unified electricity, magnetism, and light.

In a strange confluence of events, July 4th 1863 saw the preservation of the unity of the bizarre experiment in human affairs, the American Republic, as the forces of the union got twin momentous victories at Gettysburg and Vicksburg.

The effect of these victories, observations, discoveries, and new concepts greatly stimulated the creativity of the age. Jules Verne, author of many adventurous novels, published the first widely read novel depicting human space travel, *From the Earth to the Moon*, in 1865. Verne had this expedition launch from Florida in the reunified United States. In 1871, the book, *Vril: The Power of the Coming Race*, a work of science fiction, was published by an Englishman, Edward Bulwer-Lytton. It depicted human interactions with a non-human race, the Vri-ya, who predated humanity and used a powerful field called Vril by which they seemingly unified gravity and electro-magnetism to enable flight. The Vril field also enabled unheard of levels of destruction.

This idea of contact with non-human races of great power stimulated much further thought and investigation.

In 1875, Helena Blavatsky of New York City, co-founded the Theosophical Society, drawing ideas from the Vril novel.

Helena Blavatsky,1877

The Theosophical Society symbol, covering all the bases

However, the focus of creative thought returned to Mars, because of improvements in telescopes and a fascination, on Earth, with canals.

An Italian astronomer, Giovanni Schiaparelli, used the term "canali" meaning channels, to describe the streaks he observed on the surface of Mars, during one of its close approaches to Earth in 1877. This was excitedly mistranslated as "canals," by the newspapers. The Western world had already been electrified by the opening, after 10 years of construction, of the highly successful Suez Canal in 1869. Therefore, in keeping with the mental inclinations of the age, this was interpreted to mean Mars might be inhabited by intelligent beings, who like humans on Earth, built canals. This excited speculation only increased with each new "apparition" or close approach of Mars. Apparitions of Mars occurring roughly every two years and became occasions for great excitement in the press and drove astronomers to increase the power and seeing ability of new telescopes.

During that same close approach between Mars and Earth in 1877, American astronomer Asaph Hall discovered Mars had two small moons, which he named Phobos (fear) and Deimos (terror). Curiously, the same two moons of Mars appeared in Jonathan Swift's fictional novel, *Gulliver's Travels*, written in 1726. This is one of the first of many instances in history, of science fiction proving prophetic.

In 1886, Jules Verne published the novel *Robur the Conqueror*, about a Captain Nemo-like character who builds a giant, heavier-than-air, airship that terrifies the world.

Robur's Heavier-than-air ship.

France began work on the Panama Canal in 1881, but this was a far more difficult project than the Suez Canal, thus progress was very slow and the project was largely abandoned. The chief problem was the high death rate of the European workers because of malaria in jungles of Panama.

The same year, work was begun on the Corinth Canal, the dream of ancient Greece, which cut across the peninsula separating Athens from its arch enemy Sparta. This was completed in 1893. This same year a son of Boston wealth, Percival Lowell, returned from a sojourn in Japan, newly opened to the West, and a nation vigorously adopting western technology and science.

Japan Modernizes: Authoress Ichiyo Higuchi

Perhaps inspired by the exposure to a markedly different culture, whose flag depicted a great red disk, Lowell returned to the U.S. obsessed with the planet Mars and reports of its canals. Using his great wealth, earned partly as a successful author of books on Japan and Korea, he purchased a large state-of-the-art telescope and built an observatory on top of a mountain in Arizona, near Flagstaff. It was the first observatory on Earth sited to maximize good seeing conditions, and isolated from city lights.

Nikola Tesla

Also in 1893, Telsa was able to perfect his Telsa Polyphase System in the United States. This system created vortex electromagnetic fields, which, in turn made the induction motor and the AC (alternating current) electric power systems possible. Any piece of metal placed in the Tesla rotating field would rotate also, without any electrical connections to the moving metal piece being needed. Tesla then proceeded to literally *Electrify the Age.*

A Journey in Other Worlds: A Romance of the Future was a science fiction novel by John Jacob Astor IV, published in 1894. It included the concept of "Apergy," an anti-gravity energy/force that could allow interplanetary travel without rocket fuel. Astor would later perish in the sinking of the Titanic.

Spaceship flying straight up using Apergy

In November 1896, the great airship sightings began, providing sights similar to those imagined by Jules Verne in his novel *Robur the Conqueror.* The sightings moved Eastward, beginning in California and ending after a reported crash of the airship near a small town named Aurora in Texas.

The first UFO, an airship at night over Sacramento, California.

Percival Lowell, using observations from his new observatory, then the best in the world, confirmed the reports of canals on Mars, and proposed they were products of intelligence in his 1895 book *Mars.*

Percival Lowell and the Canals of Mars

In response, in 1897 H. G. Wells published *The War of the Worlds*, where Mars invades Earth using space capsules. The Martians use laser-beam weapons, and poison gas against humanity in a merciless assault that leveled cities and massacred their inhabitants, civilian and military alike. It was a vision of war on a scale of destruction and barbarity unguessed at in the Victorian age. Before the book is done, London lies in ruins, and Great Britain, and the entire world civilization by extension, is nearly destroyed. The humans respond with Army and Naval artillery, which is occasionally effective against the Martian War Machines, but is too little and too late. In general humanity collapses into a panicked mob devoid of order or will to resist. Adding to the horror, the Martians were also depicted as utilizing human blood as a food source. In general, *War of the Worlds,* in its mind-numbing depictions of mass destruction and murder set in England, of merciless war between two species of intelligence, humanity and the Martians, that had no bond in common except killer instinct, horrified its Victorian readership and changed forever the human view of the intelligent Cosmos.

Miraculously for humanity, the Martian invasion fails disastrously, because of infections by common Earth bacteria, to whom the Martians have no immunity. The novel, with is depictions of stupendous destruction, merciless struggle, and enormous loss of life, could be regarded as a prophecy of World War I and even more accurately, of World War II.

A novel entitled *Futility* was published in 1898, ending the 18th century with another vision of mass death and destruction. In the novel, an enormous newly launched ocean liner, named the Titan, and considered unsinkable, full of the rich and famous, on its maiden voyage across the Atlantic, strikes an iceberg and sinks, with almost universal loss of life among its passengers. The great ship can be thought of a metaphor for all of human civilization and its sinking, with great loss of life, as that of civilization's demise.

In what may have been the most fateful event of the end of the 19th century, in November of 1899, using the "wireless telegraphy" of Marconi, a technology based on the electro-magnetic theories of Maxwell, a message was sent 66 miles from an ocean liner to the British coast. This was one of the first long-range radio signals ever sent of Earth. This message was not just sent to England, but to the stars.

Looked upon from the perspective of the evolution of the Jungian Collective Unconscious, a common thread can be identified in the literary phenomenon of the last decade of the 19th Century. Visions of mass destruction, of merciless war with aliens from outer space, of menacing and mysterious craft in the night sky, had suddenly entered the mind of humanity. It is possible the radio experiments of Marconi had alerted some alien intelligence in the nearby stars, and Collective Human Mind saw this detection, and had recoiled from it in terror.

The 20th Century began with a scientific bombshell, destined to change the human view of physical reality. In 1900, Max Planck proposed the quantum theory of the electromagnetic field. This led to new revolution in physics. He identified a fundamental quantum of rotational momentum called "h", as residing in the electromagnetic field.

On December 17, 1903, the Wright Brothers, from Dayton Ohio made their first heavier than air flight, at Kill Devil Hills, in North Carolina. Also in 1903, Russian scientist, Konstantine Tsiolkovsky published the book *Exploration of Outer Space by Means of Rocket Devices*, which presented the mathematical theories of rocket propulsion and proposed travel by humans between planets. Humanity, as in in the Jules Vern novel *Robur the Conqueror*, had now acquired wings like the angels.

To great fanfare, America took over the Panama Canal project in 1904, completing it in 1914. This massive project created the perfect backdrop for Lowell to continue to talk about canals on Mars and publish *Mars and Its Canals* in 1906 and *Mars As the Abode of Life* in1908.

In 1905, Albert Einstein published E= mc^2 and the Special Theory of Relativity--it was the 'power of the Vril.'

In 1911, Marie Curie won the Nobel Prize in Chemistry because of her discovery of the elements Polonium and Radium, using chemical techniques she had invented for isolating radioactive isotopes. Radium demonstrated the equation of Einstein's E=mc^2 and partly solved the riddle of the power of the Sun and stars--it was from nuclear reactions of some kind. She had isolated Radium and Polonium from an ore of Uranium, called pitchblende, that was found in only one place in Europe. That place was an ethnically German region, later to become infamous, when it would become known as the Sudetenland.

In 1912, Carl Jung published his theories of the collective subconscious. Put simply, it said the human race was inherently psychic, and subconsciously knew far more about the Cosmos than it would consciously acknowledge. Also in 1912, illustrating this psychic nature of humanity, the luxury ocean liner Titanic, on its maiden voyage, ran into an iceberg and sank with tremendous loss of life. The scenario of its catastrophic sinking, reproduced in stunning detail, the sinking of the ocean liner Titan, in the fictional novel *Futility*, written 14 years previously. Nothing could confirm Jung's hypothesis of a humanity that was inherently psychic, more than this prophetic pairing of fiction and reality.

The sinking of the Titanic as in the novel Futility

In 1913, Niels Bohr, proposed his model of the hydrogen atom, the most abundant element in the universe. In an odd echo of the past, he proposed that hydrogen was like the Copernican Model of the Solar system with the heavy proton as the Sun, and the electron moving like a planet around it in several possible orbits of like those of the planets, but quantized in their orbital momentum in units of Planck's constant "h". So it was, that an understanding of the Cosmic led to an understanding of the microscopic.

In 1915, Einstein had published his General Relativity theory, proposing the bending of electromagnetic waves, light, by gravity, and with it came predictions of Black Holes, the Big Bang origin of the cosmos, and the Einstein-Rosen bridge, that might enable faster-than-light travel. The theory was verified by observations of bent starlight made during solar eclipses.

Einstein and Planck, with a portrait of Mars behind them

The tragedy of World War I followed. It claimed the life of Karl Planck, Max Planck's firstborn, at Verdun. Mixed with the horrors of trench warfare, were terrible echoes of *War of the Worlds*: the use of poison gas and, instead of the heat ray, the flamethrower. Germany was defeated, but spared occupation, and Russia experienced a Communist Revolution followed by a civil war that devastated the country. Communism won the civil war and established a dictatorship under Lenin.

In 1918, in a Germany defeated and demoralized by WWI, the Thule Society was born. It was an extension of the earlier Theosophical Society, and was filled with rich German industrialists fearful of the spread of Communism.

Symbol of the Thule Society

The society, in order to oppose communism, sponsored the Deutsche Arbeiterpartei (DAP) or German Workers' Party, which later metamorphized under its leader Adolf Hitler into the National Socialist German Workers' Party (NSDAP or Nazi Party). It was reported that members of the Nazi party became students of the occultism of the Thule Society, which included the "channeling" of messages from a race of extraterrestrials from Aldebaran, near Orion, by a woman named Maria Orsic.

Maria Orsic

Inspired by the success of Einstein's General Relativity, Kaluza, in 1919, extended the 4 dimensions of Einstein's spacetime to 5, to which Klien added in 1926, the feature of the 5th dimension being "curled up" upon itself and not visible. To the familiar four dimensions of spacetime, up-down, left and right, and forward and backward, add the dimension of time, like the fingers of one's hand, was now added a shorter fifth dimension, a thumb. Like adding a thumb to the human hand made all things possible for humanity, so the fifth dimension of Kaluza and Klein, made Cosmic reality, shaped by gravity and electro-magnetism, understandable. In the author's interpretation, the fifth dimension is the of electric charge of electron and protons. This is discussed in some detail, but aimed at a lay audience, in the book *Beyond Einstein's Unified Field: Gravity and Electro-Magnetism Redefined*. This Kaluza-Klien Theory achieved the mathematical unification of gravity and electromagnetism but raised many problems of physical plausibility. However, armed by these ideas, Einstein was encouraged as he attempted to unify the two great forces of nature, gravity and electromagnetism, in a physically testable way. Central to his attempts at field unification, however, was the rejection of quantum mechanics, and a dismissal of nuclear physics as well. Einstein concentrated on a universe containing only hydrogen (electrons and protons), which was the most abundant element in the universe. In his analysis, the other elements, and the Nuclear Forces those elements required to exist, were 'mere details.' Implicit in this unification effort was the unspoken suggestion that just as General Relativity allowed the bending of electromagnetic fields, his unification theory would allow *Electromagnetism to bend Gravity*. This possibility hovered around his unification efforts, suggesting that nullification of gravity by electromagnetism, allowing flight into space and between planets and stars, was possible without rocket fuel. This field-unification effort would occupy him the rest of his life. Tesla, not be outdone, claimed he could create antigravity with his rotating 3-phase Tesla fields. But Tesla provided no theory or experiments to back up this claim that was ever seen by the public.

Not waiting for a unified field theory, Robert Goddard journeyed to then barely-known Roswell, New Mexico, in 1930 and began launching liquid-fueled rockets, whose design was based on the theories of Tsiolkovsky, to higher and higher altitudes. His work was ignored in the United States, which was in an isolationist mood after its brutal involvement in the final years of World War I, and wanted no involvement with the rest of the Earth, much less other planets. Mars, the planet of war, was particularly worrisome. This was all understandable given the events in the world.

In an instance of science as metaphor, Subrahmanyan Chandrasekhar showed in 1931, that a burned-out star could collapse forever into a gravitational singularity. The scientific community rejected this mind-numbing possibility and came up with a host of reasons this should be impossible. Their rejection of the concept was also metaphorical, for the whole world they had known was nearing collapse into a nightmare.

Subrahmanyan Chandrasekhar, father of the "Black Hole"

Hitler became Chancellor of Germany on 30 January 1933. Einstein, being Jewish and a smart guy, as well as being a genius, understood the meaning of this fully. Einstein decided to take his family and personal secretary, who was also Jewish, on a nice trip the United States, where he had been banking money. Einstein then decided to never return to Germany.

In 1936, Otto Hahn, in Germany, assisted by his Jewish assistant Lise Meitner, began to investigate evidence that neutron bombardment of Uranium was splitting the nuclei of that element. This appeared to be a far more powerful effect than the quiet radioactivity of radium.

Lise Meitner

In 1936 , the serial movie *Flash Gorden* was released. It vividly depicted a human future of space travel in Goddard-like rocket ships, ray guns, heroes and villains. Curiously, one group of ETs, the Lion-men, were depicted as flying disc shaped spaceships rather than rockets. It also depicted a prolonged armed struggle against a powerful and evil villain, Ming the Merciless.

Flash Gordon

The looming nightmare that would become World War II was now being seen. Robert Oppenheimer, in a further demonstration of science as metaphor, confirmed the nightmarish solution of Einstein's theory of General Relativity, allowed the transformation of a brilliant star into the dark abyss of a Black Hole. It was a final singularity from which, nothing, not even light itself, could escape. It was a stunning validation and amplification of the catastrophe seen by Chandrasekhar.

In 1937, the British government approved the release of the movie *"Things to Come,"* preparing their people for World War II. In the movie, the war begins in 1940 and lasts until 1978, devastating the world. However, at the end, the world rebuilds and launches its first astronauts into space.

Making a grand synthesis all of these discoveries, new ideas, and drama, E. E. "Doc" Smith in 1937 published the epic science fiction novel, *Galactic Patrol*.

A Grand Vision of Human Destiny

In *Galactic Patrol*, the human race was a major player in a Galactic Federation of cosmic peoples, some very humanoid, others very different. They were united in their struggle against interstellar pirates, the Boskonians, who were remarkably well organized and well equipped. The novel featured beautiful women in tight spacesuits, heroic warrior astronauts, psychic communication as well as space warfare, controlled gravity, laser-like weapons, nuclear-like weapons, defector shields, and large spaceships carrying people between stars at faster-than-light speeds. It was the world of Star Wars and Star Trek, all created in the looming shadow of Mars. For the unspeakable horrors of World War II were about to break, like a monster wave, upon the whole Earth.

Lise Meitner in 1937 before she fled to Sweden

Chapter 3 The Power of the Stars

"In the second World War every bond between man and man was to perish."

Winston Churchill

Things to Come, a Movie released in 1936 by the British film industry, produced by Alexander Korda, and staring Raymond Massey, predicted a new world war by 1940 that would last until 1978. The movie, prepared by the British government, can be understood as the desperate attempt to prepare its people psychologically for war. It would be a war without pity or hope, lasting for decades, where bombs, poison gas, and specially bred plague germs were used as weapons against the civilians. The striking images of the movie were later seen in reality, in the London Blitz, and in the images of the destruction of London, echoes of *War of the Worlds.*

In March 1938, Hans Bethe proposed nuclear fusion reactions in the Sun's core, fusing Hydrogen into Helium as the source of its energy. The resulting helium 4 isotope weighed less than the hydrogen fused to make it. The missing weight, via Einstein's famous $E = mc^2$, equation, became energy, heating the Sun and the stars to incandescent heat. In the Sun and stars, gravity gathers hydrogen gas into dense clouds with hot, dense cores.

Fusion occurs there with the heat energy diffusing outward to the surface, where it makes light and heat. The same process works to fuse helium to oxygen, carbon and nitrogen in larger, hotter stars than the Sun, and onward up to iron, where nuclear fusion no longer creates energy. This solved one of the great mysteries of science at the time: "what process powered the Sun and stars, enabling all life as we knew it?"

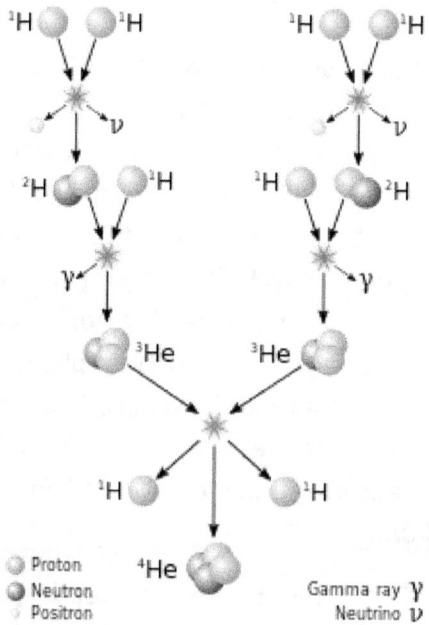

The Hydrogen Fusion reaction that powers the Sun.

Nuclear fission, the opposite process to fusion, where the nucleus of very heavy atoms splits apart releasing energy, was announced by the Kaiser Wilhelm Institute for Chemistry in December 1938. While bombarding uranium with neutrons, Otto Hahn and his colleague Fritz

Strassmann discovered that fission products such as barium were also created in the process. Hahn and Strassman also found this process released more neutrons than it consumed. A simple calculation showed that the fission process was releasing vast amounts of energy, via the relation $E = mc^2$. In January 1939, Lise Meitner and her nephew Otto Frisch provided the full physics explanation. The fission process not only released vast amounts of energy, but it created new free neutrons, and hence could be amplified into a runaway "chain reaction." Fission of uranium could thus lead to an atomic bomb. As an Austrian Jew, Meitner had emigrated from Germany in the summer of 1938, but she continued to correspond with Hahn from her exile in Sweden.

The elements heavier than iron are generally formed in very large blue stars, who having formed cores of iron explode into supernova, burning up their remaining hydrogen in one last outburst. In addition to creating a menu of heavy elements very useful to humanity, such as gold, and tungsten, the process captures some of the supernova energy into unstable superheavy elements such as uranium and thorium. Fission releases this captured supernova energy, when the nuclei split into lighter elements.

Hitler and leaders of France and England meet over the Sudetenland Crisis

The Sudetenland was desired by Germany not only for its territory, but also because a majority of its population were 'ethnically' German. That was the public reason given, at least. The truth may have been far more complex, and because of Hahn and Strassman's work, involved the stars- the Sudetenland held the only uranium mines known in Europe.

In the summer of 1938, Hitler demanded the annexation of the Sudetenland into Germany. At this point Hitler was aware that the Allies were desperate to avoid war, and thought it likely that they would appease his demands.

Hitler threatened war over the issue of the Sudetenland. On 29 – 30 September 1938, the British, Italian, French, and German leaders met in Munich to discuss the issue. The Allies agreed to concede the Sudetenland to Germany in exchange for a pledge of peace. This agreement was known as the Munich Pact. Said UK prime Minister Neville Chamberline "We have achieved peace in our time."

This claim was met with skepticism and angst in America, still in mourning for its loss of 120,000, young men, seemingly for no good purpose, in World War I. But the real subconscious reaction of America to events in Europe was revealed on Halloween Night October 30th in 1938, with a broadcast of a radio version of **War of the Worlds**. The broadcast created widespread panic, revealing another specter was haunting America. It was not just a fear of another world war, but the fear of an invasion from outer space.

Such an invasion, and merciless war that it would trigger, now seemed scientifically plausible. With Lowell's observations of possible canals on Mars, the success of Goddard's rockets-allowing space travel, and H.G. Wells horrific extrapolation of those concepts. Now was seen the fear, not acknowledged consciously, but still present none-the-less, that some morning everything humanity had built could be destroyed. America, full of strength, tested in WWI, felt invincible to any power on Earth, but what of a power not of this Earth? But dawn came the morning after Halloween in 1938, and with it the fragile dream of a world at peace was restored.

But the series of crises in Europe continued. Finally, Hitler began claiming old German territories in the newly formed nation of Poland, his neighbor. Rebuffed diplomatically from an alliance with the French and British, and fearing the Nazis, the USSR signed the Hitler-Stalin Pact with Germany on 23 August 1939. The pact was a triumph of cynical "Realpolitik": 'politics recognizing reality', buying the USSR time to prepare for the war. A war, now viewed by everyone, as the inevitable war with the Nazis.

Stalin and the Nazi Foreign Minister Ribbentrop at the Hitler-Stalin Pact signing.

Britain and France now aligned themselves with Poland. In the summer of 1939, just a year after Munich, Europe now confronted the terrifying thing that was both unthinkable and now inescapable. A Second World War. In the field of physics, something metaphorical had emerged, equally terrifying and unthinkable, and yet also recognized as physically inescapable.

It was known now that stars could burn up their supplies of lighter elements, changing them into heavier ones and producing energy up until iron, where no more energy could be produced by thermonuclear reactions. As predicted by Chandrasekhar years before, a prediction rejected as unthinkable at the time. It appeared a burned out star, if it did not explode into a Supernova, could crush itself by its own gravity until it finally no longer existed. Only a black nightmare of infinitely powerful gravity would remain to mark the star's previously brilliant life. An article written by Robert Oppenheimer, showed that the nightmare of Chandrasekhar was real and perhaps inescapable. It appeared in print on September 1, 1939. That same day, Nazi forces crossed the border with Poland, and Great Britain and France declared war on Germany two days later. This same first author of the nightmarish physics article was the same Robert Oppenheimer who would go on to lead the Manhattan Project that would end World War II six years later. Oppenheimer was thus a person who could think about the unthinkable.

Oppenheimer, Jewish and familiar with crisis engulfing Europe, was also aware of the work of Meitner, and its implications. Lise Meitner had discovered that an atomic bomb based on uranium could be built, and that Nazi Germany might be the first to build it. Oppenheimer, knowing Sanskrit, was also aware of the depictions in the of what appeared to be the use of a nuclear weapon on an ancient Hindu army armed with swords and war elephants. These are found in the translations of the *Drona Parva*, the seventh book in the *Mahabharata*.

"The valiant Aswatthaman, then, invoked the high-tier Agneya weapon, aiming at all in his vicinity. With the fiery flames of that weapon, hostile warriors fell down like trees burnt by a raging fire, a thick gloom suddenly shrouded the Pandava host, the sun was no longer visible, the very waters heated, huge elephants fell down on the earth all around uttering fierce cries, others scorched by that weapon ran hither and tither and fell down at last. Beholding the Pandava army thus burning, Kauravas filled with joy."

A vision of nuclear destruction from ancient India

The Second World War began officially 1 September 1939 and lasted until 1 September 1945. It would leave much of Europe in ruins, and killed an estimated 75 million people. It killed twice as many civilians as it did military personnel. Only in the novel *War of the Worlds*, depicting an invasion by murderous outer space aliens, was such a global catastrophe even imagined.

On October 11, 1939, Einstein, signed a letter addressed to President Franklin Rosevelt, warning him of the military implications of the discovery of nuclear fission, and of the danger of the Nazis obtaining such technology first. The letter had been written by Doctor Edward Teller and Doctor Leo Szilard, two Jewish physicists who had escaped Nazi-dominated Europe.

Edward Teller

Albert Einstein and Leo Szilard

Roosevelt read the letter and ordered a committee formed to study the matter and make recommendations.

In late April 1941, something strange happened in Missouri, the "show me state". Something crashed in the middle of the night, and reportedly a church Pastor named William Huffman was sought out to say words over the victims of the crash.

Huffman would tell his granddaughter that said the object came down in a field somewhere west of the Cape Girardeau Airport between Cape Girardeau and Chaffee. Bodies were found, and Huffman was asked to say some respectful words over them.

This seems ridiculous to some, but was in keeping with civilized custom, that the dead bodies of "people" were to receive dignified treatment. One might ask, "Why say words from the Bible over bodies of aliens from outer space?" The answer is: whose words other than the words of the God of the Whole Cosmos? Would you choose the words of Marx, or Nietzsche, instead? The fruit of that pair's teaching, it was seen even then, would soon snuff out 100 million lives. As to the official status of outer space aliens in Judeo-Christianity, this has been studied by the Catholic Jesuits, with the conclusion that outer space aliens can enjoy Christianity the same as any poor boy from Georgia[1].

When the church Pastor got there, he got the shock of his life. There was no cylindrical airplane with wings or propellers. "There was a round silver disc that was broken open, and there was metallic debris in on the ground that had set fire to the field," said Smith.

According to Huffman's granddaughter, there were three creatures at the scene. They said two had already died by the time Huffman arrived.

---

[1] The author has also studied this question and has come to the same conclusion. Detailed discussion of this is to be found in the Book *Cosmic Jesus,* from Adventures Unlimited Press.

One of them was apparently still alive, still breathing as Reverend Huffman knelt over this creature. And they were about three and a half to four feet tall. Your typical Greys as we would call them today, with big black eyes and long, thin arms and legs, and the creature expired in front of Huffman.

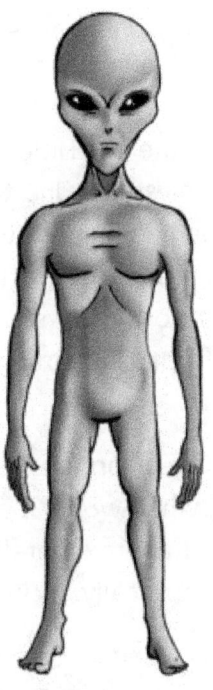

Grey Alein

Huffman said the U.S. military eventually arrived at the crash site, forced everyone to swear oaths of secrecy, and took all the evidence. Now the Federal Government knew, at its highest levels, that we were not alone in the Universe.

Harry Truman, later president, grew up in Missouri, and was high ranking Mason there. He would have known many things about this incident that were kept from the public.

Pearl Harbor was attacked December 7, 1941. Surprise was total, despite radar warning of a large formation of aircraft approaching Hawaii and intercepts and decipherment of coded Japanese messages indicating an attack was imminent. These warnings had been ignored. My father stood in a long line and joined the Army on the same day. The US declared war on Japan. However, on Dec 11 the Nazis declared war on the US. Hitler's motivations for this suicidal act remain a mystery. One theory was that Hitler believed that the Imperial Japanese would conquer the US, and leave him with no piece of North America, unless he joined in their war effort. Some also think that Hitler believed that aliens from Aldebaran, the Vril, would help him conquer the world.

In Germany, the Wanasee conference was held 20 January 1942 and organized what would be later known as the Holocaust. I would kill approximately 8 million people in Death Camps, many horrifically, including 6 million Jews.

The Imperial Japanese began to carry the war to the Continental US. On the night of February 24, a Japanese submarine surfaced off Los Angeles, California. Its crew unlimbered its deck gun, and lobbed several shells into an oil refinery on shore in Long Beach Harbor, setting a large oil storage tank ablaze. The crew then reboarded their submarine and it safely submerged by the time the armed forces in the Los Angeles area could be mobilized to look for it. Having been caught "with their pants down" the LA area remained on hair trigger alert after this. They were hoping their enemies would make another visit. They were not to be disappointed.

On the night of February 25, 1942, radar detected four flying objects, flying in the from the sea towards Los Angeles. Believing them to be Imperial Japanese bombers launched from submarines or else an aircraft carrier that had sneaked up on the West Coast, the military authorities ordered a high alert. The flying craft came in over LA seemly oblivious, or perhaps contemptuous, of the scrambled alert forces. So the objects flew in nice straight lines into a hornets nest of trigger-happy antiaircraft defenses. What followed was an epic fireworks show of search lights focusing on the objects while they were hammered by barrages of antiaircraft fire.

February 26, 1942, Los Angeles Times

It was rumored that two of the craft were brought down by anti-aircraft fire, and were recovered the next day, one in the Los Angeles forest, and another on Catalina Island. If so, nothing could be more revealing of Earth's situation vis-a-vis the Greys. Not only were the Grey ships not bullet proof, but they were also flown by beings capable of astronomical levels of stupidity.

The military authorities in LA, using a story that would become familiar, said the objects were merely weather balloons. It was a war, and people believed whatever authorities told them.

Foo fighter sightings began in the war in 1942. Both sides, the Allies and Axis, thinking they were superweapons created by the other.

Von Braun's V-2 rocket development, based on Goddard's work at Roswell, progressed. Nazi Germany would spend eventually more money on the V-2 project than the US spent on the Manhattan project. This program would send the first rockets into outer space and kill thousands of slave laborers working in underground caverns to build the rockets and thousands more in London where the rockets fell.

During this period, Max Planck visited his old friend Lise Meitner in her exile in Sweden. He told her "Horrible things are being done in Germany, and we will deserve every punishment we shall receive for them." His second-born son, Erwin, was arrested by the Gestapo in the summer of 1944 for his role in the plot on Hitler's life, and executed in October.

Lise Meitner in exile in Sweden, terror and survivor guilt in her eyes.

Inspired by Einstein's attempts to create a unified field theory of gravity and EM, a search was apparently made experimentally for EM Gravity effects, and it became apparent that coils energized by Tesla's 3-phase power would lose weight. This effect became the central focus of the most secret of all the Nazi "wonder weapon" programs, and was considered a potentially "war-winning program."

S. S. General Hans Kammler personally directed the Nazi Haunebu (discus) anti-gravity program by stirring mercury with a Telsa EM vortex. He had designed the Auschwitz Death Camp in detail, and also the underground V-2 factory at Nordhausen, where 2 slave laborers died for every V-2 produced.

S. S. General Hans Kammler

The Tesla rotating-field method reportedly produced strong antigravity, but the Telsa fields heated the mercury in the process, leading to toxic levels of mercury vapor in the subterranean and poorly ventilated chambers where the experiments were done. But by this point in the war, with thousands of men a day dying to hold back the Americans British and Canadians in the West, and the Russian Army to the East, no one cared if some scientists and engineers went mad and died. The goal was to find a technology that would miraculously win the war for the Nazis.

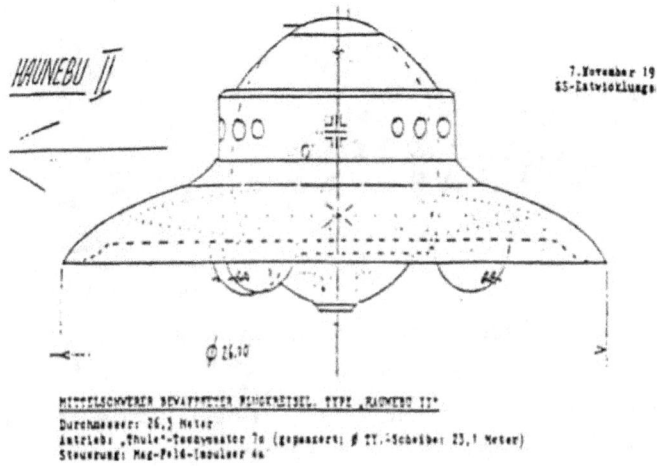

Nazi Haunebu 2 flying saucer design.

The Nazi antigravity experiments resulted in designs for flying craft that consisted of a disc shaped metal upper shell with three bumps on the bottom, which held field inductors for Tesla 3-Phase rotating power. The fact that this configuration would result in weight loss when the coils were energized, was later discovered and published by Russian and later Japanese researchers.[2]

This basic design can be derived from a basic GEM (Gravity-EM) unification theory developed by the author, and a similar theory developed by Dr. Hal Puthoff.

---

[2] The author has studied this Unified field question and has produced his own "GEM theory". Detailed discussion of this, in lay terms, is to be found in the Book *Beyond Einstein's Unified Field: Gravity & Electro-Magnetism Redefined,* available from Adventures Unlimited Press.

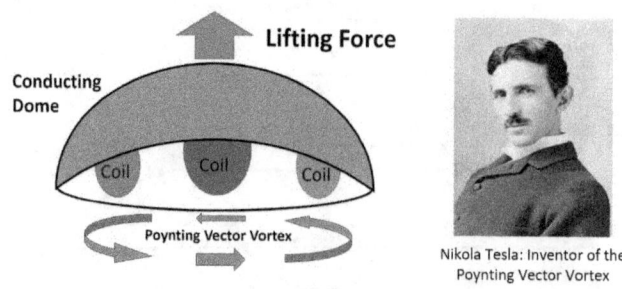

Nikola Tesla: Inventor of the
Poynting Vector Vortex

GEM Theory predicts Anti-Gravity lifting forces.

A basic design for an anti-gravity craft based on Gravity-EM unification theories of Brandenburg and also by Puthoff.

It is reported that later, in 1945, understanding the war to be lost, General Kammler had his surviving scientific staff compile an exhaustive report on the Nazi antigravity program. He then, reportedly, had his entire staff massacred by their SS guards and traded the final report to the American military for a free plane ticket to Argentina. He was never found after the war. But the Nazi antigravity program results were now in the hands of the US government.

By 1945, "foo fighter" sightings were frequent. They were reportedly only a mystifying nuisance, distracting pilots and radar operators from their duties, but doing little else.

As the Third Reich collapsed in the spring of 1945, being crushed between the American and British Imperial Forces from the West, and the Red Army moving relentlessly in from the East, the scale and savagery of the violence expanded beyond all previous imaginings. German Air defenses collapsed, and its defenseless cities were bombed and burned to rubble. The reality of the Death

Camps and the full scale of the Holocaust, long rumored in the West, was now beheld firsthand by American and British Troops, and avenged horribly by the Red Army in Eastern Germany. Nothing except the scenes in *War of the Worlds* could compare to what was being seen and done in the last days of the War in Europe.

The first nuclear weapon was detonated in 16 July 1945 in a test called *Trinity*. Witnessing the test was Oppenheimer, now the director of the Manhattan project, which had built the weapon. Beholding the test, Oppenheimer quoted the Baghavad Gita saying, "Now I am become death, the destroyer of worlds."

Harry S. Truman was now President. He was a failed haberdasher and known as a hack machine politician in Missouri. But he was also veteran of intense combat in WW I who had taken his army training in Kansas. Many were puzzled over his selection by Roosevelt as his running mate, since Truman's political career had not been impressive. However, he was a high-ranking Mason in Missouri, making him privy to many things that happened in Missouri that were not publicly known. It may be that his combat veteran status and familiarity with the Cape Girardeau incident overruled all other considerations in Franklin's mind. So, it is possible that the UFO Coverup actually began in 1937, and its deep effect on the highest circles of government grew with time, finally becoming, in the last year of WW II, a major factor in governance.

Moving quickly, the Americans dropped an atomic bomb on Hiroshima on August 6, 1945, and on Nagasaki August 9, 1945. Taken together with massive conventional fire bombing of Japanese cities and the Russian invasion of the Japanese vassal state of Manchukuo in China, the Emperor of Japan persuaded his cabinet that Japan must surrender. They feared, in particular, the possibility of a Russian Invasion of Northern Japan and the country's subsequent partitioning.

On the night of August 14/15, 1945, a Northrop P-61 Black Widow night fighter, *Lady in the Dark,* a twin-engine plane equipped with a nose radar dish and heavily armed with four 20mm cannon in its lower fuselage and four 50-caliber machine guns in a robotic turret on top of the fuselage, chased a Ki-44 "Tojo" Japanese fighter. In the darkness, with the Japanese pilot's keen eyes seeing the American plane, the Black Widow chased the Japanese plane aggressively, but was unable to lock its guns on it as it maneuvered frantically. Finally, near the ocean surface, the Ki-44 collided with a tall wave and exploded. It was considered the last air-to-air kill before World War II officially ended on September 1, 1945.

Northrop P-61 Black Widow

The Crew of the Northrop P-61 *Lady in the Dark*

So ended World War II, a seemingly endless tour of hell, where humans treated other humans as if they were invaders from outer space, and destruction and mass death equaled that depicted in the *War of the Worlds*. But it was also a time when technologies for space and weaponry leaped ahead. Was the war the Jungian collective unconscious reaction to physic contact with the Greys? The first radio signals were launched in 1900, and Foo fighters arrived in the 1942, 42 years later. This is approximately the time the radio signals from Earth would take to reach receivers in Zeta Reticule, reported to be the home system of the Greys, and 39 light years away from Earth. Let us assume that having detected these signals, they came here using "warp drive" but only to find the Earth fully armed and engaged in piteous combat, and finally armed with nuclear weapons and the capability to fly into outer space. For the Greys, it was a discouraging situation, any thought they may have had of conquering the Earth without "a shot being fired" was abandoned in the face of the astronomical levels of violence and ferocity seen on Earth.

Soviet and American Troops meet at the Elbe river in Germany.

The last laugh, Lise Meitner in England 1946

Chapter 4 Roswell and the "Watershed"

"In war, the truth is so precious, that it must go everywhere with a bodyguard of lies"

W. Churchill

With most of Europe and Asia devastated by the war, America, recovered its legions back to their homes and civilian life. But the aftershocks of the war continued. America had emerged from the war as the supreme military, scientific, and industrial power on Earth. But, as it turned out, America did secretly fear one thing above all else--an invasion by a power not of this Earth.

On March 12, 1947, President Harry Truman, announced the Truman Doctrine, aimed at opposing Russian attempts to dominate Greece and Turkey. But, while disappointing, this hardly seemed threatening to World Peace, compared to what the world had just gone through. The Russians had no nuclear weapons and no long-range bombers to carry them, even if they got them. Demobilization of the American war machine continued.

On March 1952, to its termination on December 17, 1969, Project Blue Book was run by the US Air Force. The project, headquartered at Wright-Patterson Air Force Base, Ohio, was initially directed by Captain Edward J. Ruppelt and followed projects of a similar nature such as

Project Sign, established in 1947, and Project Grudge, in 1949.

On July 4, 1947, thousands of motorcycle club riders descended on Hollister, California, and staged a drunken riot in the downtown area. The "Booze-Fighter" motorcycle club, reportedly led this disturbance. Apparently deriving their name from the term "Foo-Fighter" from World War II, this was an organization of bikers made up of restless combat veterans from WWII.

A Booze Fighter in Hollister, California.

It was a good time for publicity hounds to have a name associated with Foo-Fighters, for in late June 1947, the first great series of UFO sightings began in America. Called a "flap", a wave of UFO sightings dominated the headlines for several weeks. It started with the sighting by Kenneth

Arnold, from his airplane, of "flying saucers" while flying near Mt. Rainier in Washington State on June 24, 1947. Soon they were being seen everywhere. The "flap" continued until July 8, when it abruptly ended. What ended it?

Kenneth Arnold spots "Flying Saucers" from his plane near Mt. Rainier in Washington State.

SUNDAY, JULY 6, 1947.   SECTION A—PRICE TEN CENTS

MORE 'FLYING SAUCERS' SEEN AS MEN OF SCIENCE PONDER SERIOUS ANGLES

Disc Stops And Reverses Self In Flight, Disappears

Astronomers Say They Are Man-Made; Pilot Jokes But Wireless Report Shows Him Frightened and Nervous Later; Not Atomic—Lilienthal

By The Associated Press

The nation was baffled late Saturday night and during the day by "flying saucers" reported seen in 28 states by hundreds of persons, and conjectures came from scores of named and unnamed sources throughout the country.

Official government sources took a "let's see one" stand on the phenomenon, ad no scientists proffered a detailed explanation.

But the Los Angeles Evening Herald and Express quoted an unnamed California Institute of Technology scientist in nuclear physics as suggesting that the saucers might be the result of experiments in "transmutation of atomic energy." Dr. Harold Urey, atom scientist at the University of Chicago, called that "gibberish."

FIGHTER PLANES READY FOR DISCS

P-51 Ships Are Alerted To Photograph Objects

By The Associated Press

Col. Al Dutton, commanding officer of the Oregon National Guard at Eugene, said...

At Columbus, Ohio, Louis E. Starr, national commander-in-chief of the Veterans of Foreign Wars, asserted at a VFW convention he was expecting information from Washington about "the flurry of flying saucers." "Too little is being told to the people of this country," Starr declared.

Two Chicago astronomers said the discs are probably "man made." The indications, flashing objects "couldn't be saucers," said Dr. Gerard Kuiper, director of the University of Chicago...

Newspaper Headlines from 1947 UFO Flap.

In July 1947, on approximately the early morning of the fourth of July, while the nation was riveted with reports of the Booze-Fighter riot at Hollister, California, and elsewhere, the wave of Foo-Fighters in the skies, a cosmic event, still wrapped in mystery, apparently occurred. It would apparently, in the secret councils of the US government, produce a "watershed" moment in thinking about UFOs. This event began the UFO Coverup.

The event occurred near in Roswell New Mexico, a town already famous for the Goddard Rocket Experiments of the early 1930's. Roswell Air field was then the main base of the 509th Air Bombardment Group, the only bomb group authorized to carry nuclear weapons. Accordingly, the base, secretly, was, in 1947, the location of the entire nuclear weapon stockpile on Earth. The stockpile consisted of five plutonium "Fat Man" atomic bombs, similar to the type dropped on Nagasaki in the last days of WWII. This made Roswell the most important place on planet Earth in the eyes of the US Government.

According to Phillip Corso, the author of the book *The Day After Roswell*, the wave of UFO sightings threw the US military into a secret frenzy, and it concentrated its security assets into the Roswell area in the weeks leading up to July 4th , 1947. At its highest levels, the US government had apparently known since the Cape Girardeau, Missouri, crash in 1937 that UFOs represented possible invaders from Outer Space.

It must be understood that the then US government had just finished fighting the most massive and savage war in human history, and had just lost 300,000 dead. If the book and broadcast of *War of the Worlds* had not planted the idea of an invasion from outer space firmly in the US government's mind, then Pearl Harbor and the Holocaust had made them predisposed to form conclusions based on even fragmentary intelligence, even if those conclusions seemed initially unthinkable. If aliens from outer space captured our nuclear weapons stockpile, then Earth would be defenseless, went the line of reasoning. The probabilities of such a nightmare scenario were impossible to estimate, but the possibility could not be ruled out. Therefore, the likelihood of this possibility would be reduced by any means possible.

There is dramatic evidence that the US Government actually knew that the late June 1947 Wave of UFO sightings was going to occur. Beginning in May, 1947, they called their all-weather night fighter squadrons, equipped with P-61 Black Widows, with their WWII combat veteran crews, from their bases in Germany and Japan, back into bases in the continental US. With the Cold War with the USSR in full swing, this move made no sense. If we wanted to deter war with the Russians, why not concentrate airpower over our forces facing them? The Russians had no long-range bombers capable of reaching the Continental US, and had no nuclear weapons. They had 3 "interned" B-29s from WWII and no spare parts to keep them flying. It was not until August, 1947, that the Soviets managed to "clone" the B-29 into the Tupolev Tu-4, and so obtained an intercontinental range bomber, but this program was unknown in the US in May, 1947.

P-61 Black Widow nose radar dish exposed.

So, the stage was set for early morning July 4th, 1947, and all the actors, both terrestrial and extraterrestrial, took their places. Secrets in wartime, either Hot or Cold, are compartmentalized. Like a ship with watertight compartments, the breach of security in one area can be sealed off, saving the whole ship of secrets from sinking.

Secret knowledge is thus parceled out based on a "need to know" basis. While this "the right hand knoweth not what the left hand doeth" system preserves overall security, it predictably leads also to occasional organizational confusion in dynamic situations. In Albuquerque, and in Alamogordo, both places where fighters for air patrols would be based, Army Personnel apparently knew that the UFO wave was a possible threat to national security, whereas at Roswell, a bomber base, no one had been informed.

During the day, the latest jet fighters patrolled the skies over Roswell, and at night also fighter patrols were constant. However, the Army Air Force had no jets that could fly at night or in bad weather. Such fighters required on-board radar, which was both heavy and complicated. Therefore, the standard night fighter for the Army Air Force in July 1947, was the proven, if slow, twin engine P-61 Black Widow. They were based out of what would become Kirtland AFB in Albuquerque, to the North of Roswell, and what would become Holloman AFB near Alamogordo. Likely, the best WWII veteran P-61 crews were assigned to the task of keeping UFOs away from Roswell and its nuclear weapon stockpile.

What exactly happened on the early morning of July 4th can only be reconstructed from fragmentary reports. There was a terrible thunder storm that night, it is reported. Such storms play havoc with sensitive electronic gear and with visibility. However, the P-61 Black Widow airframes were rugged and their crews were used to operating in storm conditions from the war, especially in the Tropical Pacific. In particular, the P-61 was used to investigate violent thunderstorms after the war. The P-61 also, with their vacuum-tube radar technology and mechanical computers, shrugged off the effects of the EMP produced by nearby lightning discharges.

The same could not be said for sophisticated EM gear, likely solid state, that both guided and propelled a UFO. This is true even if fiber optics are used for data transmission, because for an EM-propelled craft, data transmitted by light must ultimately be converted into electromagnetic control fields. This is especially true if the UFO craft was flown by pilots using brain-electronics links. The brain runs on electricity, not light, so fiber optics must interface with electronics to effect commands. It is precisely these electronic/photonic link mechanisms that are most easily disrupted by EMP. Added to this, the pilots, even if used to operating on Earth since 1942, were perhaps unnerved by the violence and brilliance of the lightning storms raging near Roswell that night.

So, for the Grey aliens piloting UFOs above Roswell that night, it may have been the perfect storm of terrible weather and massive EM interference from brilliant lightning. Added to this was the fact that heavy rain conditions would absorb the same nuclear gamma radiation they were trying to monitor from the plutonium bombs stored at Roswell, as they tried to determine their number and explosive power. This would have forced the UFOs to fly closer to Roswell to get clear signals.

Based on reports of what was recovered on the ground beginning July 5, and accounts of crews of P-61s from night missions in WWII, the following sequence of events can be reconstructed. The discs waited until the storms had somewhat abated, and cruised at low altitude in a pair to gain good differential measurements of the gamma rays from the nuclear weapons stored at Roswell. This meant they were more easily seen on radar from the nearby bases and the radars in the noses of the P-61s, who converged on the disc's locations. Near the discs however, the veteran P-61 crews adopted a tactic they had used during the war. The German and Japanese late-war planes had begun to carry radar detector equipment, so they knew when the allied night fighters were gaining a lock on them. The P-61 crews had learned to turn off their radars as they closed in the for the kill, using only their highly trained night vision and special binoculars to locate and closely approach the enemy aircraft. This tactic would have been even more effective in the EMP background caused by still numerous lightning bolts. There, over Roswell in the early morning of July 4th, an epic confrontation took place.

Apparently taking the pair of discs completely by surprise while they concentrated on taking gamma ray measurements, a P-61 struck, opening fire from a 100 yards with 20mm cannons and 50 caliber machine guns. One disc apparently exploded into a cloud of debris and thus disappeared from both sight into the cloud decks below and from radar tracking. At night, it would have simply appeared to have disappeared in a flash of light. Based on previous UFO reports, the crew, temporarily blinded, probably thought the disc had darted away from them into the deep sky. The other disc was apparently intact but badly damaged by gunfire and fled away from Roswell at reduced speed with the P-61 in hot pursuit. The damaged craft was tracked by the P-61 and by radar until it reached the Plains of San Augustine, 100 miles away, roughly 15 minutes flying time for the 400 mph P-61. There, the disc crashed and a team of Army troops was sent from Alamogordo to recover it, arriving at the crash site by dawn. A leather-lunged red-haired Army Colonel was reportedly in charge of the Army ground team.

According to reports, the fairly intact disc found on the ground was made of silvery metal, and was heavily damaged and spit open, with bodies of dead aliens of the standard Grey type, both inside and outside the ship. The red-haired Colonel lined up the irrigation inspector and team of archaeologists, who had been witnesses to the recovery operation, and bellowed warnings at them that the US government would kill them and their families if they revealed what they had seen that morning.

However, the story did not end there. Four days later, Mack Brazel, a rancher living in the Roswell/Corona area, discovered a field of silver-colored metallic debris, in a pasture. He informed the Sheriff at Roswell and gave him pieces of the metallic debris. The Sheriff, in turn, informed the personnel at Roswell Army Air Base who investigated and confirmed the existence of the debris field. Given the wave of UFO sightings dominating the news media, and the obvious lack of orders as to what to do about such a development, the Base commander, Colonel Blanchard, a man well-liked by his subordinates, excitedly told the base press officer to announce "we got one!"

Roswell Daily Record Front Page July 8, 1947.

This created a minor firestorm. However, it was quickly extinguished by an angry phone call from General Ramey, in Fort Worth, Texas, who was Blanchard's superior. Soon, a red-haired, leather-lunged, Colonel arrived with a team of Army troops and took over the Roswell Army Air Base. For close to a week, the entire base was mobilized to go collect and examine the debris found near Corona. Reports were also heard of a second debris field, and the recovery of strange corpses, with small bodies, and large heads with large black eyes. Rumors also circulated that mutilated human bodies were also found, mixed with the wreckage. Everything gathered up was flown out to Carswell Army Airfield at Ft, Worth, Texas. From there, it was reportedly then taken by specially escorted ground transport to Wright Patterson Field in Ohio, where the HQ of the Air Staff was located. The base personnel were all sworn to secrecy concerning what they had seen. The special Army team then left, and things returned to a haunted normality at Roswell Army Air Base.

The story was put forth, along with sample pieces of aluminum foil and balsa wood, that the WWII veteran, highly trained, and elite members of the 509th bombing group, entrusted with the entire US nuclear weapon stockpile, were too stupid to recognize parts of a weather balloon when they examined them. Later, in the 90's, the US Air Force even claimed that reports of strange bodies recovered in 1947 were actually crash dummies dropped from planes in 1954. This suggested, seemingly, that far from being an encounter with extraterrestrials, the Roswell incident was instead an example of "time travel." This explanation for what was seen at Roswell has a surprising number of adherents even today, no doubt becoming an occasion for hilarity among the original concoctors of this tale. The fact remains that piloted space travel is a well demonstrated technology, whereas time-travel, which violates the 2nd Law of Thermodynamics, is not.

Among the explanations for the two UFO crashes reported in the Roswell incident: one into fields of metallic wreckage near Corona , indicating a mid-air explosion, and the other near Socorro, a metal disc with a gaping hole and dead crew, but largely intact, a number of causes have been proposed.

One explanation was that the craft were brought down by radar interference with their flight systems. While faintly plausible, this does not match the reports from the crash sites. Radar does not cause midair explosions or make holes in metal-skinned craft, whose structure reflects microwave radar waves. The Foo Fighters had been operating in war zones full of radar beams since 1942. To assume the discs were victims of simple radar is to assert that humans know more about building aircraft capable of flying in an intense EM environment than someone from outer space, who is supposedly far more technologically advanced than ourselves. While it is reported that intense EM pulses can bring down UFOs on occasion, these pulses are reportedly several orders of magnitude more intense than any radars beams to be found near Roswell in 1947.

The second explanation is that that the craft where brought down by lightning, in the violent thunderstorm reported at the approximate time of the crashes. But metal-skinned craft, like jet airliners, are immune to lighting, which is a phenomenon seen on both Jupiter and Venus in addition to Earth. One would have to suppose, that even after the Foo Fighters had been operating in Earth's atmosphere since 1942 that they would be susceptible to this common Earthly phenomenon. In addition, lightning does not cause sky-craft to explode in midair, unless they are full of hydrogen, like the Hindenburg. Lighting rides on the surface of metal skinned planes and exits from them to continue its course to ground, not even making holes in them. To assume the discs were victims of lightning is to once again assert that humans know more about building aircraft than someone coming here many light years from outer space.

A common, and harmless occurrence.

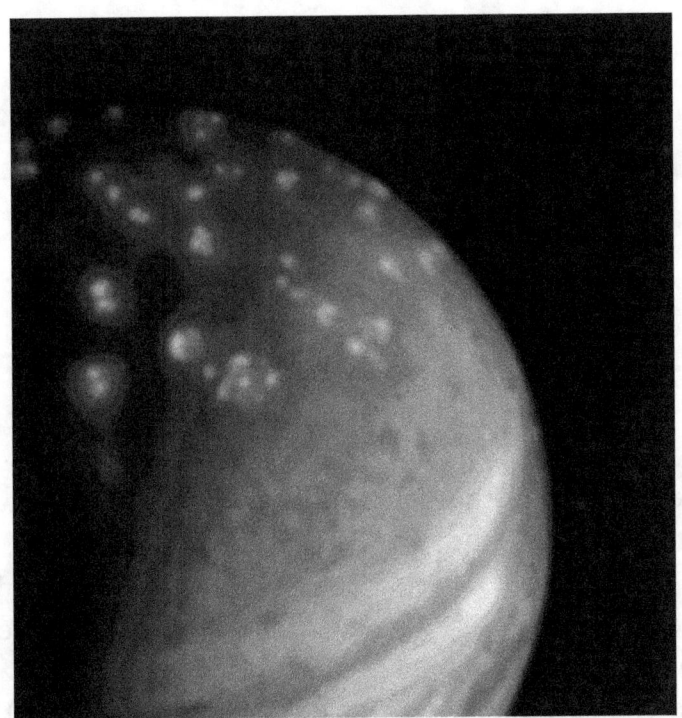

Lightning observed on Jupiter by the Juno spacecraft. NASA

Absurd, in the view of the author, is the concept that a bunch of wreckage, dead aliens feasted on by vultures and coyotes, and an intact but badly damaged UFO, full of bodies, recovered immediately by Army personnel, constituted a deliberate "gift" from the ETs. This gift was supposedly meant to promote our technological development and demonstrate the awesome superiority of the aliens. This, in the author's view, is as ridiculous as the Nazi's claiming that the Hindenburg explosion in New Jersey, was a demonstration of Nazi Air Power prior to WWII. It sounds like a tale foisted on terrified abductees, rather than an actual reality.

The Hindenburg, the pride of the Third Reich

The final explanation for the two crashes is that the two craft were brought down by gunfire from one or more P-61 Black Widows, which would have been patrolling the skies at night over Roswell to guard the US nuclear weapon stockpile. Gunfire does punch holes in aircraft, and it also makes them explode in mid-air. Given the descriptions of the debris, it did not appear to be bullet proof. So, even though the scenario of a WWII propeller-driven aircraft bringing down a spaceship from many light years away seems improbable, it also violates no law of physics. In fact, given its seeming improbability, this requires nothing more than hubris on the part of the Greys, for it to be true. To those who reject this possibility out of hand, the author responds, "Denial is not just a river in Egypt." For some, the possibility that Roswell represented a military clash, is too terrifying to contemplate. So they refuse to do it. For others the cult belief that the Greys are some invincible extraterrestrial "Godot" come to rescue the human race from itself, is so strong that details in reports of the Roswell incident no longer concern them.

P-61 Black Widow ground firing its 20-mm automatic cannons and 50-caliber machine guns.

At the morning staff meeting at the base, the Monday following the departure of the red-haired Colonel and his team, Colonel Blanchard was asked by one of his subordinates to comment on the occurrences of the previous week. Colonel Blachard's only response was to smile and chuckle, "Boy, did we screw up..."

So ended the Roswell incident, but not its effects, apparently, on thinking in the highest circles of the government. The alien attempt to investigate our nuclear weapon capability, our most powerful weapon system, apparently behavior not seen before, had convinced the higher ups that the aliens posed a grave threat to both national and world security. H.G. Wells novel, *The War of the Worlds,* became required reading. America now faced two Cold Wars, one with the USSR and the other with aliens from outer space.

Publicly, the US Air Force, now a separate service from the US Army, established Project Sign in late 1947 to study UFOs, but it was all for show; nothing about Roswell was discussed.

It is reported that in response to the Roswell Incident. President Harry Truman ordered the creation of a special committee called MJ-12 (Majestic Twelve) to study the situation and give him regular reports.

Maj. Jesse Marcel, holding supposed wreckage recovered in the Roswell Incident.

Chapter 5   From Dallas to Dulce

"He who can destroy a thing, can control a thing"

Dune, Frank Herbert

On 7 January 1948, Godman Army Airfield at Fort Knox, Kentucky, received a report from the Kentucky Highway Patrol of a UFO near Madisonville. "First the aliens were after our nukes, now they were after our gold," went the thinking deep in DC. Four P-51D Mustangs of C Flight, 165th Fighter Squadron, Kentucky Air National Guard, one piloted by Captain Thomas F. Mantell were scrambled to investigate. Mantel, a decorated WWII combat veteran pilot, was killed in this incident.

Postwar P-51 Mustang Fighters

Captain Thomas Mantell, decorated WWII combat veteran.

In response, the US Air Force phased out Project Sign, and in 1949 began Project Grudge, which at least sounded more active, but it also did little but issue reports saying UFOs were not a problem.

On August 29, 1949, the Soviets set off their first atomic bomb, a copy of the Nagasaki "Fat Man" design. The nuclear weapons arms race between the two superpowers, the US and USSR, now began. This event was probably viewed with mixed feelings within MJ-12. The nuclear weapon monopoly enjoyed by the US was over, but so was the feeling that the US would have to face an invasion from outer space alone. The US and USSR had been allies in the most massive and desperate war ever fought, and everyone in the US government, and even the USSR government remembered the days of their alliance.

Reportedly the Russians had pursued UFOs with even more vigor and ferocity than the US, in the late 40's and early 50's, under the mistaken impression that any unknown aircraft over the USSR was probably an American or British spy plane. Finally, after many lost MIG fighters, the Russians succeeded in downing a UFO near one of their top-secret military research bases. When the hatch on the crashed craft was pried open, however, they discovered the pilots were not American, but reportedly short people with big heads. This surprised the Russians, but did not make them change their tactics. The order for the Red Army and Red Air Force regarding UFOs remained the same: "shoot on sight".

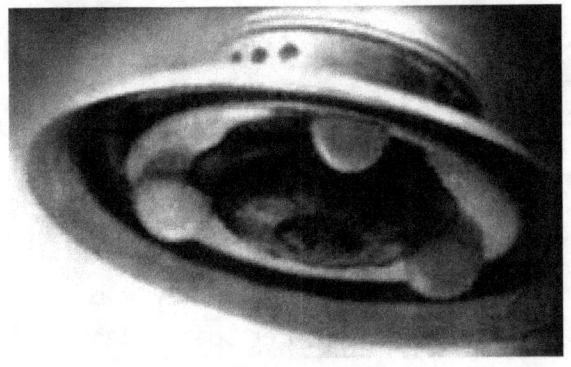

Adamski UFO photo

Hollywood had been part of the US government's WWII effort, producing both training and propaganda films, many starring John Wayne. It was now apparently enlisted in the new Cold War, and not just the Earthly one, but also the one the US government was facing in outer space. The content of these movies dealing with outer space aliens was telling.

On January 1 1951, the movie, *The Day the Earth Stood Still* was released. This movie presented the ETs as humanoid, benign, and all powerful. They had arrived in order to save us from ourselves.

The Day the Earth Stood Still

Its message was appealing, humanity was obviously in serious trouble, perhaps Outer space aliens could make things better. But this message did not last long.

The Thing from Another World

On April 6 1951 the movie, *The Thing From Another World*, was released. It portrayed the ETs as mortally hostile. A crashed flying saucer, and animal and human mutilations were seen. The alien, played by James Arness of later *Gunsmoke* fame, was depicted as posing an implacable and mortal threat to humanity. Strangely, the alien's use of human blood as a food source, seen first in the novel *War of the Worlds*, was a major plot element. The alien was killed, finally, by US Air Force personnel using a high-voltage plasma.

Taken together, the two 1951 movies *The Day* and *The Thing* could be taken as representing a debate within the government over the meaning of the UFO phenomena. However, events in the skies over Washington DC would soon end that debate.

In March, 1952, Project Grudge was replaced, like replacing a worn out used car with another worn out used car, by Project Blue Book, and it very soon became apparent that Project Blue Book was another meaningless show piece. Perhaps Grudge sounded too hostile to be a message to a public that wanted reassurance. Blue Book sounded like "business as usual" and that was the message the public needed to hear. This was especially true in the face of later events in 1952.

In July, 1952, the UFO nighttime overflights of Washington, DC, began. While publicly denying the UFOs were anything more than "thermal inversions," the US Air Force was scrambling its all-weather, radar-equipped, jet fighter, the F-94, to chase the objects. It was reported that the fighters had secret orders to shoot down the objects. This went on for weeks, until, it was later reported, a jet fighter had actually opened fire on one object and had blown a chunk off it. The piece, being recovered, was reportedly made of a curious composite of aluminum alloy and tiny ceramic spheres. The overflights stopped after this classified incident. Project Blue Book basically made no report on the incidents: "nothing here to see, folks."

Looked at in retrospect, the UFO overflights of the US Capitol could be easily interpreted as an attempt to intimidate the US government. Such displays of air power above cities to induce "shock and awe" have many examples in human geopolitics, and have been quite effective at times. A recent example of this was the use of December 1989 overflights by US F-4 fighters over Manila to intimidate elements of the Philippines military that were attempting a coup against then-President Cory Aquino. The tactic was successful and the rebel troops returned to their barracks.

The US government's Hollywood story machine was now thrown into even stronger action. Hollywood was now firmly hostile to the aliens. In August 1953, the movie, *War of the Worlds*, a lavishly produced technicolor epic was released. Like the novel, the aliens in the movie were depicted as mass murdering implacable foes of human civilization, laying waste to cities, and using advanced technologies that made their machines invincible to human weaponry. Not even nuclear weapons were effective against them. Once again, as in the original novel, Earth was saved by the lack of alien immunity to common Earth bacteria.

Reported UFO Image from 1952 in New Jersey

In November 1953 an F-89 Scorpion, radar equipped, jet fighter, piloted by Lt. Felix Eugene Moncla Jr. was lost over Lake Superior while doing a night-intercept of a UFO. The radar blips of the fighter and the UFO were seen to merge on the radar scopes of nearby Kinross Air Force base, but then all contact with the fighter was lost and it was assumed to have crashed into the lake. Neither the fighter nor the body of its pilot were ever recovered.

The F-89 Scorpion radar equipped fighter

Lt. Felix Eugene Moncla Jr.

Hollywood followed again with two black-and white Grade-B movies: *Them*, in 1954, which depicted UFOs as giant flying ants, also depicted human mutilations. This was followed in 1956 by the Classic *Earth Versus the Flying Saucers*, depicting human-launched space satellites and genetically deficient aliens who looked like Greys. In this latter movie, high-powered EM beams were used to disrupt alien antigravity drives, causing them to crash.

The new President was Eisenhower, an experienced war leader.

On November 18, 1952, President-elect Eisenhower received a briefing on document MJ 12. He was now "up to speed" regarding UFOs. In a mysterious incident in Southern California on February 20, 1954, he reportedly met with a party of very human looking aliens at Edwards AFB. They reportedly asked him to give up all his nuclear weapons, a request he politely refused. How could he give them up when then Nikita Khrushchev now had them? The meeting reportedly ended shortly after this.

Later that year, in a Flag Day speech on June 14, 1954, Eisenhower discussed why he had put "under God" in the pledge of allegiance:

"In this way we are reaffirming the transcendence of religious faith in America's heritage and future; in this way we shall constantly strengthen those spiritual weapons which forever will be our country's most powerful resource in peace and war."

He also got the phrase "In God We Trust" put on all US currency. As a WWII commander, he knew the US now needed a powerful ally, and he was reaching out the most powerful Ally possible. Soon it was plainly seen why Eisenhower was seized by this impulse.

Sergeant Jonathan P. Lovette, serving at White Sands, NM, White Sands Missile Range (WSMR) was reportedly abducted in March, 1956, by a flying saucer in broad daylight, and his body was found days later, ten miles away and horribly mutilated. In particular, his eyes and tongue had been removed surgically, as were his sex organs and rectum. Further, his body was completely drained of blood, yet with no collapse of the veins and arteries. This effect had never been seen, except perhaps in Nazi death camp experiments. His death was listed as unsolved.

To WWII combat veterans, atrocities of this type were familiar, it was only the source of the atrocities that was puzzling. However, atrocities are common in warfare because they are part of the "Psy-Op" aspect of war, whose target is the morale of one's enemies. This meant the Greys were apparently waging the most brutal type of psychological warfare against the US government.

In 1956 the movie *Forbidden Planet* was released. It featured humanity in the future journeying to a haunted Mars-like planet using a faster-than-light driven, saucer shaped craft. A sci-fi action thriller but also a deep psychological inquiry into the nature of evil, the movie posited that evil existed even among supposedly much more advanced peoples than humanity.

Forbidden Planet, **with our very own flying saucer.**

On October 4, 1957, another "Psy-Op" occurred--this one far more benign. The USSR launched the first Earth satellite, "Sputnik", triggering the "Space Race" between the US and the USSR. This led to an explosion of high technology research in the US. By orbiting the Earth with an object, the USSR had showed that it, temporarily at least, controlled the "high ground" and could project power anywhere on the Earth's surface. The US, using the genius of Von Braun and his fellow German rocket scientists, was soon to catch up to this accomplishment.

In December, 1957, a remarkable technicolor movie called the *Mysterians*, produced by the same Japanese studios as the Godzilla movies, was released. It depicted abductions of women into flying saucers, hybridization experiments, underground alien bases, alien bases on the far side of the Moon, saucer overflights of Tokyo to create shock and awe, human traitors intoxicated by the sight of advanced alien technologies, genetically damaged aliens, and, with a bow to Godzilla, a giant alien robot destroying buildings. The aliens in the movie showed utter disregard for human life, either military or civilian. The aliens were defeated by human mastery of directed-energy weapons, which included a relativistic electron-beam weapon. Great entertainment for the author, later, at 12 years old, as a Saturday matinee movie.

The *Mysterians* , from the folks who brought you *Godzilla*

In 1959, between Feb 1 and 2, a party of nine Russian snow hikers disappeared near a place called Dyatlov Pass in the Northern Ural Mountains. Search parties found their tent, covered with light snow, but no bodies inside it. Instead, they found the tent had been sliced open from the inside and the 9 people inside had fled in apparent terror from it. They had fled, many without shoes or even socks, into the -30 F degree snowstorm in the night. By following footprints, Soviet authorities found six bodies, all frozen to death. The missing three other hikers were not found until a month later. They had not died of cold, but from massive crushing injuries, like from a car accident. They had all been mutilated and were missing eyes, tongues, and in one case, eyebrows. One of the bodies was radioactive. No further details were provided in the reports. Local people reported orange orbs of light had been visible in the sky during the nights of the hikers' deaths. This, like the Sergeant Lovette case, was apparently a message being sent to the governments of the US and USSR. However, the message was peculiar, indicating not only utterly psychotic cruelty, but also weakness, because the people killed and mutilated were attacked in isolated areas far from any help or rescue. It could be interpreted as "payback" for the "shoot on sight" orders now being implemented against UFOs by both governments.

The Victims of the Dyatlov Pass Incident

UFOs photographed near the Pantex nuclear weapons fabrication facility , late 50's

On 12 April, 1961, Yuri Gargarin, a Soviet fighter-pilot-turned "cosmonaut", became the first human being to go into orbit. He was followed shortly thereafter into space by Alan Shepard of the US, and later still by John Glenn into orbit. The Space Race was on.

Yuri Gargarin

John Glenn, First American in Orbit

President John F. Kennedy assumed the office of President on January 1961. Then, on April 17, 1961, the abortive Bay of Pigs invasion was launched and failed. It was a bad way to start a Presidency.

On May 12, 1962, General Douglas MacArthur, gave his famous "Duty, Honor, Country" speech at West Point, which included these words:

"We are reaching out for a new and boundless frontier.

We speak in strange terms: of harnessing the cosmic

energy; of making winds and tides work for us; of creating

unheard synthetic materials to supplement or even

replace our old standard basics; to purify sea water for our

drink; of mining ocean floors for new fields of wealth and

food; of disease preventatives to expand life into the

hundreds of years; of controlling the weather for a more

equitable distribution of heat and cold, of rain and shine; of space ships to the moon; of the primary target in war, no longer limited to the armed forces of an enemy, but instead to include his civil populations; of ultimate conflict between a united human race and the sinister forces of some other planetary galaxy; of such dreams and fantasies as to make life the most exciting of all time."

However, life would become more exciting than people anticipated. On October 14, 1962, CIA spy planes photographed Soviet nuclear capable missiles in Cuba, triggering the Cuban Missile Crisis. It was resolved peacefully and the Russians withdrew out their missiles. However, the world had experienced a close encounter with nuclear war. Nuclear weapon stockpiles were growing enormously, and for the first time in history, it appeared the human race had become capable of turning the Earth into a radioactive wasteland--a seemingly dubious accomplishment--however, it was perhaps also a useful one, given the secret realities of the moment. Following the Cuban missile crisis Kennedy attempted to go forward positively.

April 5, 1962, *Moon Pilot*, by Walt Disney pictures was released, with Tom Tryon as an astronaut and Dany Saval as Lyrae. Everyone should watch it.

Lyrae is an alien woman, a Pleiadean, with brown eyes and auburn hair, who saves an astronaut's life, played by Tyron. She claims to be from Beta Lyra, the 2nd brightest star in the constellation Lyra, the Harp. They fall in love and later have children together. It was unique in that it presented the concept of very humanoid aliens, basically cousins, of humans on Earth. The movie helped generate public support for the Moon program.

Moon Pilot, from Walt Disney, with a Pleiadean on the far left

In order to further advance US prestige in the world, Kennedy declared the goal of placing a man on the Moon and returning him safely to Earth on Sept 12 ,1962, saying :

"No nation which expects to be the leader of other nations can expect to stay behind in this race for space. ... We choose to go to the Moon in this decade and do the other things, not because they are easy, but because they are hard."

Project Blue Book, the public Air Force program to study UFOs stuttered along. It was basically a "hood ornament" on a much vaster and more secretive program focused on UFOs.

Dallas, November 22, 1963, John F. Kennedy was assassinated. Much has been written about this event, and will be written. JFK was a man of deep personal flaws, and had many enemies. It is possible he even was responsible for the death of Marylyn Monroe, who he had apparently had an affair with. He had also antagonized major organized crime figures and people connected with the CIA/Cuban exile community. However, a recent suggestion has also been made that he wanted to end the UFO coverup and join in an alliance with the Russians against the Greys. A speech reportedly announcing these goals was apparently slated to be delivered in Dallas the same day he was assassinated. Accordingly, it is possible that he was the most high-profile victim of the ruthless cult of secrecy surrounding the UFO Coverup. In any case, his Presidency passed to Lyndon Johnson of Texas. Johnson was an ardent proponent of human space exploration.

Johnson also inherited the American role in Vietnam, which became a bleeding wound on his Presidency. While militarily our armed forces did well in Vietnam, domestic support for the war began to falter. It appeared that the US was losing this aspect of the Cold War.

In the mid-60's UFO sightings continued with large surge in sightings in 1966. Witness's reports of a saucer landing in Michigan with a corrugated metal hull were dismissed by Blue Book as "swamp gas". This confirmed what many had gathered already, that Blue Book existed to manage public opinion about UFOs and not to investigate them.

Images of moving UFOs taken during the "swamp gas" UFO surge.

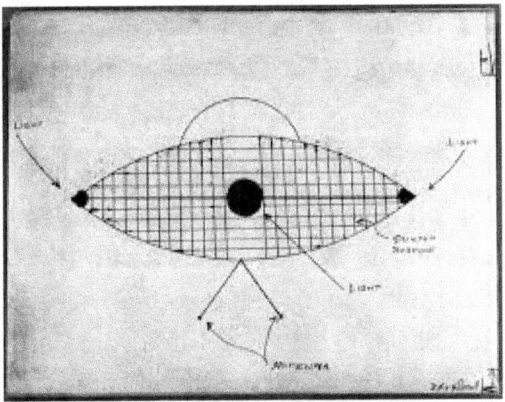

A drawing by a witness of a saucer that landed during the "swamp gas" surge.

Threatened both on Earth and from space, the US government reportedly sought to stabilize the outer space front. The US reportedly reached out to the Greys to make a deal with them: shoot on sight would be abandoned, and instead peaceful scientific exchange would occur. The Greys would not block our access to the Moon, and in return we would allow them to gather genetic data from the US population without interference.

The Betty and Barney Hill case was publicized in the book *The interrupted Journey: Two Lost hours Aboard a Flying Saucer,* by John Fuller, on January 1, 1966. This documented evidence for a UFO abduction that occurred in 1961 and had been repressed in the memories of the man and wife who had been abducted. The book portrayed the abduction as basically harmless even if disconcerting.

Secretly, sometime near the Johnson-Nixon Presidential transition in January 1969, a face-to-face meeting supposedly occurred between US government representatives and the Grey aliens at Holloman Air Force Base near White Sands, New Mexico. This was the event apparently celebrated in the Spielberg movie of 1975: *Close Encounters of the Third Kind*

It is reported that a "Secret Treaty" with the Grey aliens was signed at the Holloman meeting. The treaty was intended to be a treaty of friendship and scientific exchange: The Greys would provide us with space technology and not interfere with our space program, and in return they would be granted the right to abduct a small number of designated American citizens, to take samples of sperm and eggs from them to learn about our genetic makeup. The Secret Treaty was a complete breach of constitutional processes: the treaty was negotiated without the "advice and consent" of the US Senate, by ambassadors it did not appoint, and the US Senate was never allowed to debate and ratify the treaty. In the defense of this deed, authorities who negotiated it insisted, it was an act of cosmic "Realpolitik", stabilizing the Outer Space front, even while the US position on Earth suffered increasing damage because of the war in Vietnam. "Desperate Times Require Measures" was the operational phrase of the moment.

With the "shoot on sight" orders lifted, UFO encounters became much more relaxed. It was the age of "Cosmic Détente". A great deal of knowledge was gained from many face-to-face meetings with many alien species besides the Greys. It was discovered, as would be expected scientifically, given that the human race was a relative newcomer to the Cosmos, that a vast "zoo" of different alien species existed and were coming here, with apparently, a vast array of intentions. At least 50 different species were identified, with scientific curiosity being the main motivator for their visits here, to observe the Earth and its children. It became obvious from this that Earth existed in some sort of "Romulan Neutral Zone" in interstellar space, that was controlled by no one and where all species could freely travel.

The television series *Star Trek*, provided a vision for the human race in the future as being an honored and valuable member of the "Federation" of like-minded species, enjoying "Warp Drive" for faster-than-light travel, and the opportunity to explore and settle many faraway worlds orbiting other stars. It was a revival of the vision of the human race of the future, first portrayed in "Doc" Smith's *Galactic Patrol* in 1937. All that was missing was the planet *Trenco*.

Original Starship Enterprise: our own saucer

Two important premises were features in the *Star Trek* series, which ran from 1966 to 1969. One was the "Prime Directive" forbidding contact between advanced races in the Federation and less advanced species. Primitive species must be allowed to "find their own way" into the interstellar community, and only be contacted directly after achieving a crucial set of cultural and scientific milestones.

The second premise, less often mentioned, was the frequent occurrence of basically human ETs on various faraway worlds, as if the stars had been "seeded" long ago with the human gene-spore. The Vulcans , Romulans, and Klingons were prime examples of this phenomena, which was acknowledged in the series as a "mystery", but allowed for green dancing girls to occasionally appear and beguile Captain Kirk. All in all, *Star Trek* presented a vision of hope for the human future in the Cosmos.

It was learned that while the Cosmos was a complex tapestry like Earth, however, three basic genotypes dominated the UFO alien encounter reports. A dominant presence was the so-called Greys, reportedly, resembling social insects in both behavior and physical characteristics, with sexless workers, and sexually equipped higher "drone and queen" castes. The Greys reportedly owed their color to the fact that their oxygen-carrying blood pigment was based on copper, not iron, and is blue, like that found in Horseshoe crabs and Octopi. Next most commonly reported, were remarkably human-like aliens termed variously Lyraens or Pleiadeans, also called "Nordics". They were reportedly, over all, quite friendly to human beings, and seem to regard us as kindred. Finally, repto-avian aliens, "Draconians" have been reported, looking like tailless lizard-men and reportedly 7-feet tall. They reportedly came from stars in the constellation Draco--the Dragon. This type was chiefly reported from encounters with US troops in the DMZ area of Vietnam. It was reported that the Pleiadeans did not get along with the Greys, and had even worse relations with the Draconians, whom they reported had chased them out of their original home in the Constellation Lyra, forcing them to relocate to the Pleiades. Thus, the Earth and its troubles appeared to be a reflection of the troubled Cosmos, rather than an aberration on it.

Seemingly in a public consequence to the secret accord reached at Holloman AFB, the Air Force declared that it was going out of the UFO investigation business. The infamous Condon Report was issued in 1968, basically declaring UFOs as simply a case of mass hysteria and misidentification. Project Blue Book was officially ended in December 1969. After this, if you spotted a UFO, there was no government office to report it to. The UFO Community responded by forming the Mutual UFO Network, or MUFON, to receive and record UFO reports.

But a new aspect of the UFO phenomenon had emerged during this period. On September 7, 1967, Snippy, a horse, was found by its owners, horribly mutilated and with all its blood drained, but with no vascular collapse as a result. Such a thing was never seen before, except perhaps in Nazi Death Camp Experiments. No tracks were found around the body and a large UFO had been reported on the day Snippy was found.

Snippy the Horse in death

But the Snippy case was considered a minor, if disturbing detail: even the Condon report acknowledged it as a "bizarre" UFO incident. But for those in the government, the Holloman Secret Treaty signing was considered the arrival of the new millennium. The Moon landings succeeded, beginning in July 1969, giving the US an enormous boost in morale as its war effort in Vietnam failed on the home front. The war was ended, at least in terms of American involvement. As a follow up, Henry Kissinger embarked on an energetic campaign of Détente' on Earth.

By the 1970's the nuclear arsenals of both the US and USSR exceeded that required to turn the Earth into a radioactive wasteland by many times over. It was also known to be possible to trigger nuclear weapons without any electronic controls, since the final action on a nuclear detonation was simple explosives. So, if the goal was merely to destroy the Earth as a habitable planet, this goal was easy to achieve with a disciplined chain of command, and could not be stopped once ordered. This was important because another Secret Treaty existed, reportedly, but this was one was between human beings.

In a secret understanding or treaty, according to Phillip Corso, that predated the Holloman agreement, the US and the USSR had agreed to fight as allies, as in WWII, if the Earth was invaded from outer space. After the Sargent Lovette and Dyatlov Pass incidents, the US and USSR finally had agreed on one thing: "Victory or Death" in the case of an invasion from Outer Space. This agreement also stipulated that if the Earth was about to fall, all nuclear weapons were to be detonated in place. Seeing as the ground detonation of nuclear weapons creates maximum radioactive fallout, this would require no missile launches to destroy the Earth as an abode of life. It was like wiring your house with dynamite, to be lit by a simple fuse with a match, in order to prevent its seizure by the bank. This suited the Russians just fine, since their own military continued its "shoot on sight" policy regarding UFOs. To the Russians, Dyatlov Pass meant far more than promises from the Greys.

In 1976 the US Space Program achieved an epic triumph when it launched the Viking Probes to Mars, and the two large probes assumed orbit around the planet. They also carried landers, one on each orbiter, to achieve a soft touchdown on Mars, take pictures, sample the atmosphere, and test the soil for life. One of the target sites for a lander was a place called Cydonia. Images were taken of the site area using the orbiter's cameras, and they revealed what appeared to be giant carved humanoid face, and nearby, a 5-sided pyramid. The lander was diverted from this site and the image was dismissed as "trick of light and shadow". The landers landed successfully, but when they tested the soil for life, it appeared they gave a positive signal at both sites. This result was immediately dismissed as "weird chemistry." However, the real shock came when the landers assayed the isotopic makeup of the Martian atmosphere.

The Face and nearby Pyramid at Cydonia Mensa on Mars

The isotopic signatures found by the landers indicated that Mars had been the site of a massive thermonuclear holocaust. This horrifying result, and the images found, are discussed in detail in the author's book *Death On Mars*[3].

The year 1976 ended with a Mars coverup being appended on to the already-existing UFO coverup.

In America, things bumbled along. Animal mutilations, often associated with UFO sightings, became a chronic hazard of livestock ranching in many areas of the American West. This attracted the interest of a remarkable television reporter, Linda Moulton Howe, who quickly realized that the Government official explanation--predators or vultures--was simply not true. She documented this in her excellent book *An Alien Harvest*. The parallels between the pattern of mutilations and those seen in both the Sgt. Lovette case and the Dyatlov Pass incident were not mentioned, as being too deeply terrifying to make a good news story. However, it was becoming glaringly apparent that the Hollomon Accords with the Greys were not being followed.

Many more people were being abducted, than were listed in information provided by the Greys, and some of them were never returned alive. Animal and occasional human mutilations were becoming frequent. To make matters worse, the technological sharing of information was not yielding anything of real value. At the secret bases jointly occupied by the Grey aliens and the US military, tension began to build. Finally, one day in 1978, according to numerous reports, during the Presidency of Jimmy Carter, something finally snapped.

---

[3] Available from Adventures Unlimited Press.

President Jimmy Carter, a one-term Governor of Georgia, and a man of deep religious faith, assumed office on January 20, 1977.

President Jimmy Carter

President Carter, having witnessed a UFO in 1969, along with several other people, was eager, upon assuming office, to find out what the highest levels of the US government actually knew about them. But he quickly found out he was being stonewalled by whomever controlled that knowledge. No one would tell him anything, and NASA, the Pentagon, and the intelligence agencies answered his inquires with only bland smiles. The truth was, after the disastrous Watergate Scandal, and Dallas before that, the "powers that be" in the UFO realm decided that the Office of the Presidency was too unstable and transitory to entrust with UFO secrets. Carter was left to fume at the shocking realization that a secret US government existed that did not answer to anyone. So it was with astonishment that he was one day confronted with the ghastly truths of the UFO situation.

According to several reports, Carter was in the Oval Office one day in late 1978, when a party of high-ranking military and intelligence agency officials arrived to see him urgently. It seems that near the town of Dulce, New Mexico, under nearby Archuleta Mesa, a secret base jointly occupied by the US military and Grey aliens was located. At this base, human captives were being held, and were being subjected to horrific experiments. Pressed beyond all endurance, a sizable body of US special forces, had attempted to rescue the human captives. This action resulted in a major firefight with the aliens. Accounts vary as to the number of those killed, human and alien, in this horrific subterranean battle. It seems certain that more than 50 humans, captive and special forces were killed, together with a smaller number of aliens. The Secret Treaty regime had collapsed.

It is reported that President Carter wept openly when told of this battle, yet recovered himself and ordered that "shoot on sight" orders be reinstituted against the Greys. No doubt, his deep religious faith was his firm strength, now that he realized everyone he had trusted as President had been lying to him. This order was complicated by the fact that some UFOs were regarded as belonging to "friendlies." But the Greys dominated reports of alien activities at Earth, so the order was fairly easy to implement. So the great "New Millenium" of ET relations ended in an episode of horrific violence.

In June 1979, apparently in response to both *Close Encounters of the Third Kind*, and the secret clash at Dulce in 1978, the graphic sci-fi horror movie *Alien* was released. Its themes were the idea of an alien life form that was the ultimate Darwinian "survivor", but more important was the secondary theme of utter betrayal of human beings by other human beings. For in the movie, the "powers that be" had chosen to sacrifice the lives of the spaceship crew for the "cause of science."

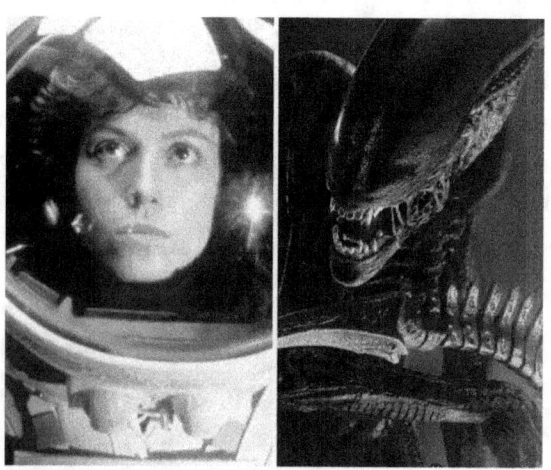

Sigourney Weaver and the main protagonist from *Alien*

Chapter 6 Star Wars to Star Trek

"Location, location, location..."

Old saying the real estate industry.

"It's so easy to keep secrets that nobody wants to know..."

Cassandra Chen, Morningstar Pass the Collapse of the UFO Coverup

President Ronald Reagan assumed office January 20, 1980. Because of his consulting with a group headed by Edward Teller , the father of the US Hydrogen Bomb and the founder of Lawrence Livermore National Laboratory, a center of nuclear weapon and fusion energy research, Reagan entered office probably more cognizant of the UFO situation than any president in history. Building on the arms buildup begun by Jimmy Carter in the last years of his presidency, Reagan expanded and intensified the massive arms buildup with emphasis on space weapons. The spectacular success of the massive US space shuttle, a Buck Rodgers spaceship compared to the primitive and tiny Soyuz manned spacecraft of the Russians, made the Star Wars space defense program seem imminently doable.

The Russians, behind the US technologically, and now locked in a grinding war with Muslim fundamentalists in Afghanistan, had become completely dependent on sales of petroleum from their oil fields in Siberia to finance their empire. However, the price of oil was $80 US a barrel, more than enough to cover Russian production costs, so everything was fine for the Soviets. Reagan was about to change that.

For the Saudi Arabians, two enemies existed. One was the Godless Soviets, fighting their fellow Muslims in Afghanistan. The other, even more terrifying enemy, was the US Fusion Energy Program. Fusion, the power of the stars, was the only proposed energy source that could safely replace Fossil Fuels. Fusion reactors, could run day or night, in fair weather or foul, and had a limitless fuel source found in every glass of water. They could not melt down like existing Fission Nuclear reactors, produced no long-lived nuclear waste, and could not have their fuel rods diverted to make nuclear weapons. Fusion was the power source of the future as surely as the Sun rose in the morning and stars came out at night. The US Fusion Energy Program had been heavily funded by President Jimmy Carter, a trained nuclear engineer. It had been making rapid progress in the late 1980's. It was the Saudi's worst nightmare.

In order to win the Cold War, Reagan offered the Saudi's a "deal". He would cut the US fusion research budget, and they would respond by pumping as much oil as possible out of their sunny desert oil fields. The Saudi's would create an "Oil Glut," dropping the price of oil far below the production costs of the USSR, who had to pump oil out of the frozen ground in the long Siberian Winter. This is exactly what transpired. The US cut its fusion research budget in 1985 and 1986, and the Saudi's responded by creating the Oil Glut in 1986, cutting the market price of oil from above $80 a barrel to below $40 a barrel. The USSR soon got a note from its bank, telling them their accounts were overdrawn. Added to the disastrous war in Afghanistan, the Soviets began to make gestures of peace.

Reagan, encouraged by Margret Thatcher of Great Britain, responded nobly to the overtures of Mikhail Gorbachev, the new Soviet leader. In the back of Reagan's mind, it seemed, was the certainty that the USSR was a valuable ally against his real enemy, the Greys. The Russian were smart, well-armed, and they could fight fearlessly. As the USSR faltered, Reagan offered the strong hand of America to bear them up. Reagan began to make a series of public statements, which paraphrased, said the following: "If the world was threatened by an invasion from outer space, would we not all set aside our differences and fight together to save Humanity?" He made these statements on four different occasions. He was not only speaking to the cognizant in the US and USSR, but to the Greys themselves. Soon, with astonishing speed, the Cold War faded in 1989.

Curiously, triangular craft began appearing frequently in UFO reports during this period, preserving the "Tesla-

Three Phase Coils" design indicated by GEM unification theories, but with the disc-like dome absent or trimmed back to reduce weight. It has been speculated that the triangle craft may be our own technological emulation of the ET antigravity craft, trotted out of hiding to "muddy the waters" concerning real ET UFO sightings.

Triangle UFO

Consistent with the formation of a covert, global anti-Grey alliance, reports of extensive intelligence sharing with the Soviets were then reported. At Wright-Patterson Air Force Base in Ohio, reportedly the US repository of an enormous quantity of captured UFO hardware, it was said that teams of Russian and even Communist Chinese military officers were flown in and given extensive tours and briefings. Reports were also heard of several battles in the sky over Russia between UFOs of different types. The human race under Reagan, had formed a Rainbow Alliance against the Greys and anyone else with designs on this planet. Shoot downs of Grey discs increased, using bullets, missiles, and EMP weapons. This was one way to lower the number of abductions, which now had become a chronic problem. The nuclear stockpiles of both the US and USSR were reduced, but the warheads were not dismantled, but instead put in safe storage, "in case of a rainy day."

Reagan and Gorbachev: Why did the Cold War end so quickly and peacefully?

It must be understood that modern hydrogen bombs are triple action. They are initiated by a small fission bomb, roughly the yield of the Hiroshima weapon. This then ignites the fusion reaction. The high-energy neutrons put out by the fusion reactions then trigger fission reactions in a casing enclosing the fusion portion. These final fisson reactions, from normal uranium found in ore, double the explosive yield of the hydrogen bomb and create almost all its nuclear fallout. Thus, a modern hydrogen bomb is a fission-fusion-fission device. The Hydrogen bomb yield can be further boosted by simply packing more raw uranium around it. This increases both the explosive yield and its radioactive fallout. That means, if your goal is simply to set off all your hydrogen bombs on the ground, in order "deny the use of the Earth" to somebody, you can make this even more certain by packing any uranium you have handy around the nuclear weapons you possess. Those weapons, in turn, whose initiation begins with high explosives, can actually be lit by an ordinary fuse, like on any firecracker.

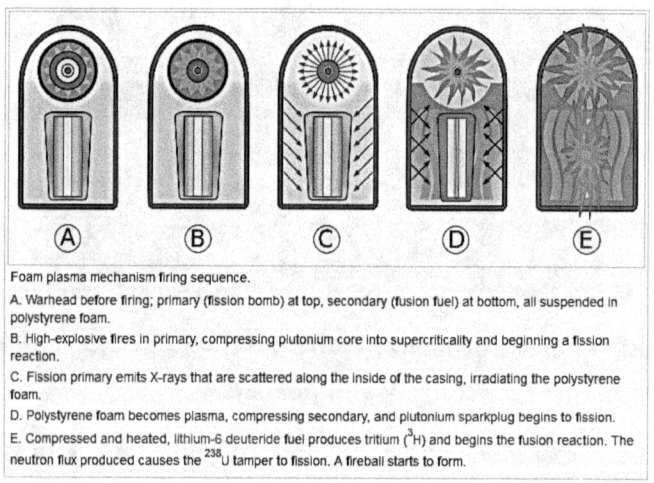

Foam plasma mechanism firing sequence.

A. Warhead before firing; primary (fission bomb) at top, secondary (fusion fuel) at bottom, all suspended in polystyrene foam.

B. High-explosive fires in primary, compressing plutonium core into supercriticality and beginning a fission reaction.

C. Fission primary emits X-rays that are scattered along the inside of the casing, irradiating the polystyrene foam.

D. Polystyrene foam becomes plasma, compressing secondary, and plutonium sparkplug begins to fission.

E. Compressed and heated, lithium-6 deuteride fuel produces tritium (^{3}H) and begins the fusion reaction. The neutron flux produced causes the ^{238}U tamper to fission. A fireball starts to form.

Hydrogen Bomb "triple action" (Wikipedia)

The "trim weights" used in military and even civilian aircraft control surfaces were switched from lead to depleted uranium in the 1980's. On one level this makes no sense, since depleted uranium, uranium 238, from which all uranium 235 had been removed (0.3%), is a gamma-ray emitter, making such trim weights detectable at a distance and a radiation hazard in case of an aircraft fire or crash. Uranium metal in these trim weights is flammable and slightly radioactive. Tungsten is denser and of roughly equal abundance, is much cheaper, and is not a fire or radiation hazard compared to uranium. This strange substitution was done, apparently, so that "on a rainy day" the uranium weights could be taken from the aircraft and packed around ground-based thermonuclear weapons, in order to increase both their explosive yield and ability to create long-lived fallout. The same use could be made of widely used depleted uranium slugs used in US anti-tank rounds. Reagan was saying to the Greys and Draconians, apparently, "you want this planet, come and take it!"

Looked at from afar, with a sea of stars behind us, the Earth is a beautiful blue planet, rich in that elixir of life, water. The basis for all known life is the chemicals hydrogen, oxygen, nitrogen, and carbon. These are, with inert helium, the most abundant elements everywhere in the universe. This makes Earth life a sample of most life elsewhere, and certainly describes the chemistry of the recovered bodies of aliens from UFO crashes. It also explains why some species would look upon this planet, , in the words of H.G. Wells, in *War of the Worlds,* with "Envious Eyes." However, it has these pesky humans living on it and they are heavily armed with nuclear weapons--enough weapons to turn Earth into a radioactive twin of Mars. Humans are also known to fight to the death against invaders. Both these factors, it can be argued, have helped deter any thought of an invasion from outer space from covetous neighbors.

So what, besides claiming to abductees that they could take much better care of Earth than we can, are the Grey's to do? An outright invasion will simply trigger a nuclear war, ruining the intended prize of the whole operation with terrible losses to their own side as well. Such a move could also trigger an interstellar war, if the Earth is in a strategic location.

Taking a page from Sun Tzu, the master of war in ancient China, one must conduct a war of stealth and treachery whenever possible. Sun Tzu said the best way to win a war is without any fighting at all. Practice "formlessness", keeping your enemy guessing continually about your true intentions. Then overwhelm them with "shock and awe" so they will lose all will to fight. The enemy surrenders peacefully. "No muss, no fuss" is Sun Tzu's ideal war.

The concept that ET craft are invincible is belied by many reports of crashes.

A careful analysis of Grey intentions over the decades has revealed the following:

Grey intentions are, in the coming decades, that Grey-Human hybrid beings will be the only form of technological intelligence inhabiting this planet, human beings being too unpredictable and dangerous to allow a shared existence. Hybrid beings are to be used, because using human sperm and eggs recovered in abductions, the human immune system against common bacteria of Earth can be transferred to the Grey colonizers. This colonization is to be accomplished through stealth, intimidation, and terror, so that open war can be avoided. This hypothesis was published in the MUFON Journal in 1992 as *A Hypothesis of Reticulin Intentions.* Reportedly, it created alarm and dismay, even within the UFO Cover-up leadership

Ironically, the UFO cover-up, meant to help defend humanity, helps the Greys in their intentioned strategy, by making the UFO subject a continued matter of mystery and fear. One cannot not logically analyze a situation, if people one trust continually deny it exists. Therefore, it can be said that Disclosure equals Survival for humanity.

A similar estimate of Grey alien intentions has been arrived at by Dr. David Jacobs , after working extensively with the Grey alien abductee community, and is summarized in his 1998 book *The Threat.*

As an added complexity, it is widely reported that Pleiadean type aliens are here living among us, undetected. They are apparently producing children with terrestrial humans. However, in the author's opinion, the genetic kinship of terrestrial humans and Pleiadean type ETs makes it inappropriate to call such children "hybrids" as in the case of the Greys. The are simply the products of human intermarriage.

Hollywood, after brief infatuation with the scenario "the Grey aliens are here to save us" concept in the 70's and early 1980's, apparently was brought up to speed by the agencies after Dulce, and went back on track. *Independence Day*, a movie released in 1996, portrayed a *War of the Worlds* situation with the aliens depicted as mass-murdering fiends.

The Movie Independence Day

Finally, even Steven Spielberg, of *Close Encounters of the 3rd Kind*, in the 1970's and *ET the Extraterrestrial* , in the early 1980's, investigated the whole matter. He then released the excellent and horrific *War of the Worlds,* with Tom Cruise, in 2005. True to the original novel of H.G. Wells, it portrayed the aliens as mass-murdering blood-suckers.

Sorry friends, the Cosmos is the same screwed up place as the Earth is. Indeed, they both are part of the same fabric.

War of the Worlds 2005

Taken as a whole, it is now possible to arrive at a Situation Estimate for Humanity and its place in the Cosmos, with an eye towards predictions for our future.

The Phoenix lights

It is apparent from both reports of many different alien cultures visiting here, and maps of the stellar neighborhoods of the Solar System, that we probably exist in a neutral zone in interstellar space and we are located close to its strategic center. Also apparent is the fact that planets like Earth, with moderate gravity and abundant liquid water, are rare jewels in the Cosmos, since only a few have been found of the thousands of extrasolar planets now identified, Added to this, is our strategic location in the galactic neighborhood, at almost at the center of the Orion Spur, a bridge of stars that connects the major Sagittarius and Cygnus arms of the Galaxy. This means, that in the Cosmic game of Monopoly, we have issued Boardwalk and Park Place, in the initial parceling out of properties to begin the game. It also probably accounts for why Humanity, despite or lower state of technology, has been allowed to keep this fine piece of Cosmic real estate. For one of our advanced neighboring species to try to forcibly take this planet would probably trigger an interstellar war with other surrounding powers.

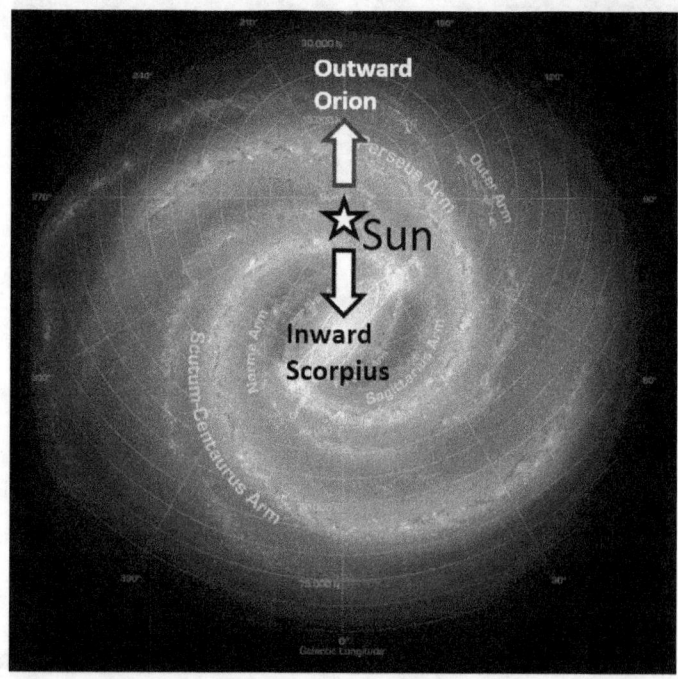

The location of Earth viewed from above the Galactic plane

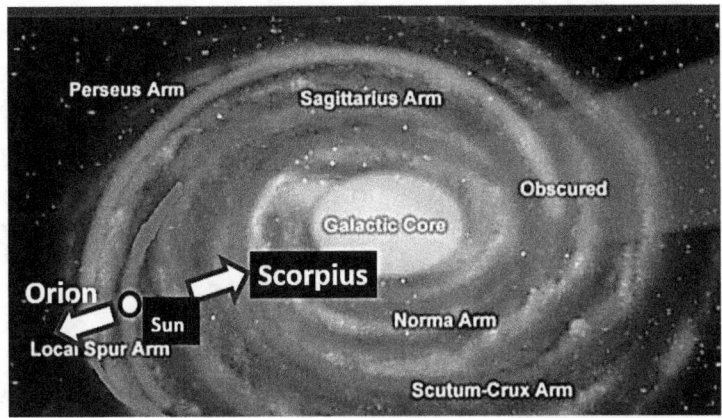

Our strategic location in the middle of the Local Spur Arm

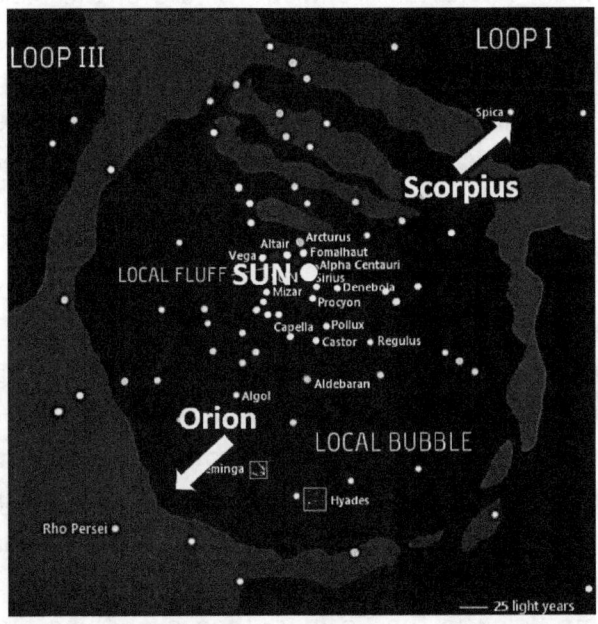

Earth and The Solar System appear to be in the center of a strategic "Local Bubble" in the Galaxy

We can be likened to the country of Luxembourg in Europe, a small weak country bordered by Belgium, an ally of Britain in the West, France to the South, and Germany to the East. It survived and remained fully sovereign precisely because it was bordered on all sides by major military powers.

So far so good for the Human race. We got lucky. However, despite our being located in the center of a neutral zone, we have apparently been born into a multi-millennium interstellar conflict. We are of the Pleiadean genotype, and this genotype has reportedly been in mortal conflict with the Draconian, or repo-avian, genotype for many generations. This is easy to see in our folklore. The story of our creation "in God's image", and the expulsion from the Garden of Eden because of a serpent may be garbled versions of our seeding here by very humanoid aliens and the reports of the Pleiadeans being driven out of the original home in Lyra by the Draconians. So also, the western mythological tales of brave heroes fighting dragons and the association of the dragon with the Devil in Revelations amplify this hostility towards intelligent and powerful reptiles. Nothing that the author is saying here should be construed as diminishing the Bible as source of spiritual truth. For if one believes in a God who is the Supreme Being in the Cosmos, then everything that has occurred is according to His will, and the rest is details. Also, one would expect many levels of meaning in any communication from a Supreme Being to us. So, we were apparently born into an interstellar conflict that has run hot and cold for millennia.

Saint George slays the Dragon

The Grey aliens are apparently newcomers, like ourselves, to this cosmic situation, and also do not get along with the Pleiadeans. It is possible they are simply mercenaries employed by the Draconians or someone else unknown.

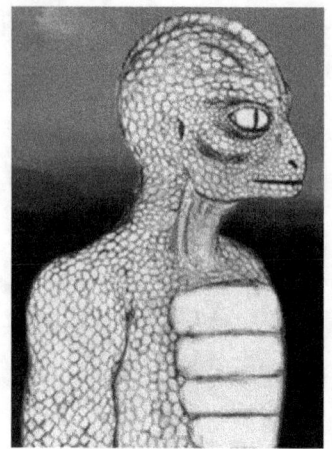

Draconian

Grey

Pleiadean (Lyrae from Moon Pilot)

Like any analysis based on fragmentary knowledge, this Pleiadean/Draconian/Grey basic genotype model is probably grossly oversimplified. There are probably alien types who do not fit into any the three categories; Mantis-like beings for example. There are probably some types of Greys who are friendly and respectful of human beings. There are probably also Pleiadeans who view Earth humans with great hostility, especially since they would covet our nice planet. However, one must start somewhere in trying to make sense of our situation, and this three-genotype model agrees with most of what is reported. It must also be noted as a sign of further good fortune, given the fearsome reputation of Dragons in our folklore, that Draconians are only rarely reported, and have not reportedly ever been recovered from UFO crashes.

Finally, it must be noted that this concept of a Cosmos full of intelligent beings dominated by three basic genotypes and also full of conflicts between them, is supported by a completely independent line of evidence from Mars. Details on this discovery can be found in the author's book *Death on Mars*, published by Adventures Unlimited Press, and in the documentary *Blue Planet Red* by Brian Dobbs.

On Mars, large humanoid faces are found carved near pyramids like those at Giza. The faces are humanoid, even allowing for erosion, and do not resemble Greys or Dragons yet are countless millennia in age, indicating humanity far predates its Earthly manifestation. Sadly, the humanoid civilization on Mars was utterly destroyed by a massive thermonuclear attack from space[4]. This shows not only do human beings predate their appearance on Earth, but so do the tragedies that have affected our species here.

Evidence of Dead, Humanoid Indigenous Civilization on Mars
Objects imaged at Cydonia Mensa and Galaxias Chaos

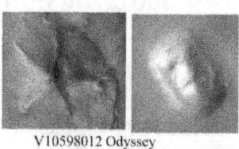

V10598012 Odyssey
Cydonia Mensa

Archeology on Mars is very ancient "from before Dinosaurs"

V22286011 Odyssey
Galaxias Chaos

Humanoid Presence in Solar System is Long Standing

Summary of Humanoid Archeology found on Mars

In Summary, therefore, the Earth and Humanity are apparently in the situation we are now in because of the

[4] The isotopic evidence and NASA-gathered images supporting this discovery are discussed in the book *Death on Mars*, by the author, from Adventures Unlimited Press.

Roswell incident and its aftermath. It can be said that in the crucial correlation between heavy UFO activity and location of the world's total nuclear weapon stockpile, that occurred at Roswell, the US government completely lost its innocence regarding the intentions of the UFO aliens towards humanity. The aliens, the US government concluded, are hostile and seek to conquer this planet. Hence, their interest in our ultimate means of defense. They also discovered that the UFOs were not invincible, a fact apparently confirmed many times since 1947. Everything that has happened since then, the numerous reports of UFO interference with nuclear weapons, the animal/human mutilation phenomena, reported military clashes, and finally the attempted creation of an alien/human hybrid race to replace humanity on this fertile planet, has confirmed the conclusion made after Roswell. As a result, the human race has been fighting a Cold War with the Greys ever since Roswell. This has been a contest of intelligence and counter intelligence, employing people and methods that are ugly but familiar in human-human affairs. In particular, it has required the US and other governments to lie continually to their people. However, this effort has been successful by some measures, most importantly in deterring an invasion from outer space.

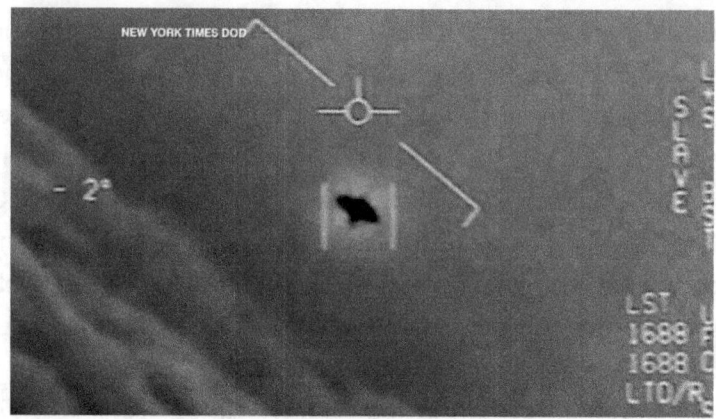

Official Navy image of UFO

More importantly, the UFO Coverup has also protected the human image of itself as masters of the only part of the Cosmos that is important. When Roswell occurred, we had no knowledge of how to fly in space, and no knowledge of planets outside our own solar system. Now we enjoy knowledges of both. This means that being told we are not alone in the Cosmos is no longer a deeply shocking statement. To this statement must also be added, that just like Earth, the Cosmos is home of both good and evil.

Based on these successes of the UFO Coverup, and the fact it was carried out by people mostly just following orders, this author would recommend a blanket pardon be extended to all those Coverup personnel who will cooperate fully in Disclosure. The Greys are our enemies, not our fellow human beings.

That said, the Coverup has had its dubious, and bad, and reportedly horrific, days. Its recent dubious moment came in an attempt at Orwellian Newspeak, to rename UFOs to UAPs. This renaming was apparently a semi-humorous attempt to transform all "Ufologists" into "U-apologists." However, the author dismisses this new term and advises the UFO community to do likewise. The government can rename UFOs once it identifies them for us, not before.

It is reported that some witnesses have been driven insane or even killed, if they did not listen to government warnings to be silent. Unfortunately, finding out who performed theses things or authorized them will be nearly impossible, given that the UFO Coverup apparatus existed to preserve secrets, not keep records. These darker aspects of the UFO Coverup and the alien presence it keeps secret are discussed in the noir science fiction novel, *Morningstar Pass, the Collapse of the UFO Coverup*, published in 2005. The novel was written by the author under the pen-name "Victor Norgarde"- after experiencing 9-11 as part of the defense and intelligence establishment in Washington DC. In the novel two brave women television journalists, bring down the UFO Coverup. The "Stonewall" Coverup dies a horrific, violent death in this novel. The Collapse is catastrophic, and leads to a desperate attempted coup de tat' by the secret government, aimed at preserving the Coverup. Sadly, in the novel, the Coverup nearly survives because much of the public is too terrified by the truth to face it.

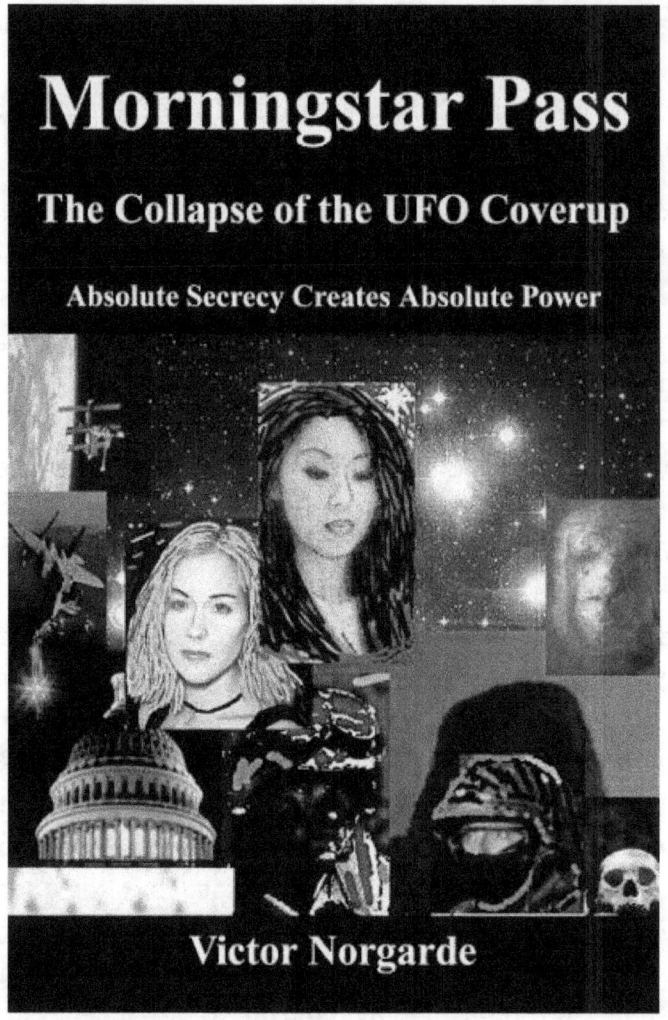

Morningstar Pass, the Collapse of the UFO Coverup

Failing orderly Disclosure by the government, the next best thing is Exposure, the open discussion of this threat to Humanity, and the means to defend against it. It is the fondest hope of the author, that the Greys will see the unity and determination of the people in the face of these knowledges, and decide to leave this area peacefully. It is the also the fondest hope and belief of the author, that the human race, with its own desperate ingenuity, its valor and ferocity, together with help from our kindred in the stars, and our ultimate source of help, God Himself, will get through this difficult transition and ultimately take our place in an enlightened community of peoples in the stars. We will eventually become part of the Federation seen in Star Trek.

Our Destiny

Epilogue: Independence Day

At dawn on Independence Day, a solitary P-61 Black Widow glided in for a landing at Kirtland Army Air Field in the cloudy sky. So, it was reported. As it taxied off the runway a flight of P-80 Shooting Star jet fighters was taking off to form the daylight patrol to the Southeast. The jet fighters thundered South in the now brightening sky, towards White Sands, where one of the captured V-2 missiles would be launched later that day. The V-2 were prizes of war recovered from Nazi Germany, along with Werner Von Braun and his fellow scientists. They now formed a cornerstone of the most advanced military research in the US. The atomic bomb was now a demonstrated reality, and soon to be demonstrated would be the ability to deliver one, via outer space, to a target hundreds of miles away.

A solitary P-61 returns to base.

Waiting for the P-61's crew to disembark, was a party of grim-faced Army Air Force Officers. The pilot of the Black Widow knew from the faces of the officers waiting for them, that their fears were confirmed: the two missing P-61s were lost, with their crews. The officers immediately took the exhausted and hungry crew away to be debriefed, where the crew learned that the missing P-61s had "disappeared from radar" over the San Andreas Mountains to the South, which bordered the closely guarded White Sand Missile Range. They had vanished after being vectored in by Alamogordo Army Air Field to intercept an unknown airborne target seen on the base's radar. They had all been operating under strict radio communication discipline during their night patrol, and so the missing planes were a mystery to the surviving crew, one of many terrible mysteries that had unfolded that previous night.

Only later, after their extensive debrief, meals, showers, and a few hours of troubled sleep, did the haunted surviving crew gather to stoically toast their fallen comrades. The wreckage of the other planes and the bodies of their crews had now been found in the desert mountains overlooking White Sands and the Trinity Nuclear Test Site, a place where the old world had ended, and a new world had been born to replace it. Being given a night to rest, the crew watched the 4th of July fireworks blossom over Albuquerque in varying states of alcoholic stupor.   In the hearts of everyman of the Black Widow crew, the sight and sound of the fireworks prompted a mixture of relief, sorrow, and grim satisfaction. They had met the terror by night and conquered it. They had been assigned a mission, and they had accomplished it. In doing so they knew they had helped save the world, even if it was now secretly transformed.

The surviving Black Widow crew from that fateful night patrol were commended and received decorations for their service and were rotated back to Hamilton field the next day. They were also told that the incident they had been involved in was considered Top Secret and they could not speak of it to anyone.

Afterward

"The Sleeper Must Awaken"

Dune, Frank Herbert

One can be a hard-working tradesman, or an accountant, or an insurance salesman, and a loving father to one's children and to one's wife. Then one night a fire breaks out in your house and you must become a different person; a dynamic, heroic and quick-thinking, person to rescue your family from danger. Similarly, my father, on the day of Pearl Harbor, was transformed from a hard-working college student, home for Christmas break, into an eager warrior for his country, as he stood in line, a line that ran around the block several times, to join the Army in Bismarck, North Dakota.

Lt. John T. Brandenburg USAAC , 8th Air Force

He flew thirty combat missions over occupied Europe, and 80% of his bombing group, the 492nd, was shot down. In the euphemism of the times the 492nd was called a "hard luck group." The story of his war experiences is told in the book *Letters from Bud*, edited by Molly Brandenburg. In his own words, he had survived the war because of "dumb luck". He later returned home to the US, married a wonderful woman and raised a family of four children. So, he survived the ordeal of combat, and later led a fruitful and prosperous life as an MD. One can therefore confront great change when it is forced upon you, endure the transition it requires, and then prosper in the new Cosmos, that one finds oneself in after that transition.

Dr. John and Muriel Ann Brandenburg and children, all finally sitting still

In a similar way, we must now awaken a person who sleeps within us, full of talents and strengths unguessed, because a time of change has come.

In retrospect, it is now obvious why the UFO Coverup which really began at Roswell, in 1947, has endured. It is explained by the Jungian Collective Unconscious, which has informed humanity all along what it faced in the stars. There is good, but also evil, in the Heavens. There is the promise of fellowship but also mortal conflict. The human race has known this, at a deep level since roughly 1900.

The collision at Roswell only reinforced what was already known and seen at a glimpse in the reaction to the *War of the Worlds* Broadcast of 1938. But now we must face our fears consciously and clearly. The UFO Coverup must end and we must reckon with the living Cosmos around us. We are, by Divine Providence, relatively secure in the Cosmos, at this time at least. Our continued sovereignty of this planet, despite its failings, proves this. Between the benefits of our location, the geo-politics of the surrounding Cosmos, our enormous stockpile of nuclear weapons, and reputation for valor and mindless ferocity, we will continue to be sovereign. What were once vices are now virtues. But we cannot merely continue business as usual. The UFO Coverup must end.

The public must be consciously acclimated to the realities of the Cosmos. Hollywood has a great role to play in this regard. The author suggests that a movie version of *Morningstar Pass* would be very helpful. The scientific community must overcome its paralysis in the face of mounting evidence of biology elsewhere in the Cosmos, as well as the physics and technologies to transit it at faster-than-light speeds. In particular, String Theory, with its countless new hidden dimensions, attempting to create a "theory of everything" must be set aside in favor of something resembling Einstein Unification, "a theory of what is actually seen." That effort seems to only require one new hidden 5th dimension. There will time to explore the meaning of the 11 other hidden dimensions at a later date, once the urgent business of understanding how people from many light years distance seem to travel here easily, and, once here, seem to fly around without needing rocket fuel or seat belts.

Now is the Winter of our Discontent. Gaia, the Greek name for Mother Nature, is angry with us because we have abused the gifts She has given us, in our beautiful home. The Human race also faces two epic confrontations with NHI (non-human intelligence), one from outer space and another created by our own hands: AI. Both have the potential to annihilate humanity as we have known it since its origins. However, Gaia can be wooed, and mollified by our adoption of fusion power to replace fossil fuels. Like the alligator and the bald eagle, nature will recover marvelously once given respite. The AI threat to the human future is mostly hype at this point, a "bright, shiny object" of Wall Street that constitutes more a threat to human livelihoods than human lives, at this time. We will have a day of reckoning between AI and humanity later when the main problem facing us has been dealt with.

Humanity, in the author's view, now faces, in the problem of the Greys, a critical moment. It is this problem which forces us to rapidly end the UFO Coverup, to allow efficient mobilization against it. We must face things unflinchingly that would have filled us with terror yesterday. We must invent things that we would have only dreamed of the night before and boldly do things we have only dreamed of the day before. Humanity must now become a new people, braver, stronger, more intelligent than we ever were before, in order to deal with the newly discovered realities of the Cosmos. We will have help from our kindred people in the stars. We will have the help of

God Himself. We will get through this transition and enjoy a world similar to that depicted in Star Trek. Behold the Pleiadeans, who could pass for any one of us on Earth, yet they fly powerful starships. We can learn their ways and become their allies.

In the word of the apostle Paul :

"When I was a child, *I spoke as a* child, *I understood as a* child, I thought as a child*: but* when *I became a man, I put away childish things* "

So, the Sleeper must awaken in all of us, and our future will be bright, as will the future of our children, and their children after them.

The sleeper must awaken.

The process we are about to go through is similar to graduating from high school, where you had mastered the skills of young adulthood, and become a high school senior, an object of awe and veneration by your underclassmen. With your high school diploma in hand, now you go to a new city, exchanging a dorm-room for a bedroom, and must make new friends and master new rules of behavior. You are no longer a senior, but now are again a lowly freshman. The world has turned upside down. You have to take all sorts of new courses, and learn many new things that "they did not teach you in high school", both in and out of class. Some of the things you learn now are deeply troubling, other things deeply enlightening. You have to also figure out what you are going to do for a living once you get out of college, and choose your courses of study accordingly. You have to become a real adult now.

Whole societies have gone through this process successfully, most notably the Japanese. The collision with the Spanish and their attempts to conquer Japan, either with conquistadors or missionaries, made Japan retreat into isolation. The arrival of Commodore Perry's Naval squadron, 200 years later, was deliberately intimidating, but also peaceful. Its arrival caused a revolution in Japanese thought and understanding of its place in the larger world. Despite some "What the Hell?" apocalyptic displays in the streets, where the veneer of civilization seemed to completely unravel in some villages as word of Perry's visit spread, and the immediate suppression of such nonsense, by the Samurai lords, the eventual reaction of the educated Japanese elite was an insatiable curiosity about the larger world they now found themselves in. Japan embarked on a deeply serious and organized program to learn the ways, especially military technologies, of the larger world. Learned people were sent abroad to study, foreign experts and technologies were imported and learned from. This program was very successful, with Japan becoming a shining light to all the less developed peoples of the world. That is, Japan engaged the wider world it found itself in, rather than hiding from it. We must likewise, as a society, engage with the living Cosmos. Our kindred in the stars, it is reported, will help us in this process.

So it can be done, and, as for the mistakes of some societies in this transition, such as the rise of Imperial

Japanese militarism, we have our opportunity not to repeat them. In the end, with God's help, and with good will and eagerness to learn from our neighbors, we will achieve the vision of Star Trek.

A model of Engagement :The Best and Brightest of Japan bravely go to study at Western Universities

A day to look forward to in the future, from the movie
Star Trek: First Contact

One final mystery remains: why is the Earth and its Children such an object of fascination for its neighboring peoples, at least 50 types who have visited here, mostly motivated, reportedly, by pure curiosity? Why is this speck of dust, out of a Universe full of specks of dust, so attention getting?

As a scientist, I must turn to where all people of learning turn when their learning can provide no answers. It is God, the Creator and Sustainer of the Cosmos, who knows all mysteries, including this one. Our only clue is that the world translated as "World" in the New Testament is actually the translation of the Greek word "Cosmos", so that the verse, familiar to many, actually reads, in Modern Speech, " For God so loved the Cosmos, that he gave His only begotten Son, that whomever believeth on Him will not perish but have eternal life." Or elsewhere, where John the Baptist exclaims over Jesus, "Behold the Lamb of God, who takes away the sins of the Cosmos!" Christ was born in a stable here, on this speck of dust, in an oppressed corner of a mighty Empire, was crucified and rose again, defying death, so it is written.

Is it possible, we are an object of such fascination for our interstellar neighbors, because Earth is actually the most important planet in God's Universe?

The author's, response to this question: ASTONISHMENT.